SUNLIGHT ON CLIMATE CHANGE

A Heretic's Guide to Global Climate Hysteria

Ron Barmby
" Dare to Know ! "

RONALD PAUL BARMBY

Tellwell Talent
www.tellwell.ca

ISBN
978-0-2288-3134-1 (Hardcover)
978-0-2288-3133-4 (Paperback)
978-0-2288-3135-8 (eBook)

Quaecumque Vera

(Whatsoever Things Are True)

For Lysle

"You are the finest, loveliest, tenderest, and most beautiful person I have ever known—and even that is an understatement."

– F. Scott Fitzgerald

Table of Contents

Preface ix
The climate change debate needs diversity, facts and fairness.

An Introduction to Global Climate Hysteria 1
There is a difference between a Heretic and a Denier. "Dare to know! Have the courage to use your own understanding."

Chapter 1: Carbon Dioxide Deserves a Fair Trial 9
- The Prosecution's case against Carbon Dioxide:
  CO2 is an industrial pollutant that will cause runaway global temperature increases and kill the oceans.

- The Defence's case for Carbon Dioxide:
  CO2 is essential to all life in all its forms, its effect as a greenhouse gas is minimal, and its impact on the oceans and marine life is essentially nil.

Chapter 2: The Sun is the Smoking Gun of Climate Change 41
- The Sun is the Gun; Sunspots Are the Smoke:
  Sunspots have been used for hundreds of years to successfully predict climate.

- The Earth is a Moving Target: Milankovitch Cycles and the Milky Way Rotation:
  Let's consider the ice ages and why they happened.

Chapter 3: History is the First Casualty **56**
The global warming community keeps shortening the definition of recorded history to avoid well documented events that clearly demonstrate climate change occurs naturally.

Chapter 4: There is No Climate Emergency; There is a Crisis in the Intergovernmental Panel on Climate Change **71**
- The Scientific Method vs. The IPCC Method:
  Are 25% of your friends Elvis Presley impersonators?

- A Heretic's Interpretation – There Is No Climate Emergency:
  Forget the forecasts; did global warming stop in 1998?

- The Crisis in the IPCC:
  The Iron Lady helped establish the IPCC, then turned against it. Since then, it has got much worse.

Chapter 5: The Good, The Bad, and The Ugliest CO2 Reduction Ideas **107**
Here are five green initiatives that were not thought all the way through. Give the bill to the taxpayer, the environment, or the very poor.

Chapter 6: Quaecumque Vera (Whatsoever Things Are True) **126**
Ten things you should know that are true. #1 is there never was a 97% consensus among scientists that man-made global warming is real and dangerous.

Chapter 7: Saving the Planet............................ 151

The 2015 Paris Agreement's current failure rate is 99.99%. There is a better way.

Acknowledgements.. 165

List of Figures 167

Bibliography.. 169

About the Author 171

Preface

The basis of the theory that human emissions of carbon dioxide contribute to the greenhouse gas effect is undeniable. However, the current international debate on the impact of greenhouse gases on global warming lacks a diversity of viewpoints. Non-conformists get shouted out by every level of media and our politicians ridicule dissenters. The result is public policy on climate change that is wrong and dangerous to our society because fact-based decisions have been replaced with fear-based decisions.

Governments have adopted climate change policies to appease activists and lobbyists rather than cope effectively with climate change. While the world maintains its focus on greenhouse gas emissions from environmentally conscious economies, there are major economies that are polluting the air, water and land without restraint or accountability. That is not fair, and worse, it is not wise.

I am a climate change heretic, and I disagree with academics that have enabled climate change science to be taken over by politics. They have exaggerated the dangers and simplified the causes.

We all share this planet and have a right to join the debate on climate change based on truth. And when we find the truth we need the courage to share it.

That is why I wrote this book.

An Introduction to Global Climate Hysteria

I'm a climate change heretic, not a climate change denier. A heretic rejects an established faith-based doctrine, whereas a denier rejects an established set of reality-based facts. A heretic does not accept an idea when proof of it does not exist; a denier does not accept an idea even when proof of it does exist. A heretic stands on principle; a denier stands on prejudice.

This is an important distinction for a member of the worldwide profession of engineering, which builds the infrastructure that is the foundation of our advanced civilization, the industries that employ countless millions, and the machines that convert yesterday's science fiction into today's reality.

The job of Professional Engineers requires a highly advanced education in the laws of science, and the ability to apply that knowledge to the construction of things that have practical uses. It is this combination of the theoretical and the practical that differentiates us from the purely academic scientists. Engineers, by training, know the same advanced laws of thermodynamics, chemistry and physics as do research scientists, and we are great at mathematics. Still, we know from experience that you cannot push on a rope. I belong to a disciplinary sliver of engineering that utilizes the earth sciences, including a detailed history of the biosphere, the thin layer of life on the surface of our planet, and how it has changed over time and is still changing. This is the training and experience I bring to the issue of our changing climate.

As a heretic, I don't deny facts that are established by the scientific method. I won't deny that humans have created many environmental disasters (I acknowledge the role my profession has played in this) and I don't deny that our current exploitation of the Earth's natural resources and generation of pollution must change before we ruin it all. The climate is changing—it always has been—but as we will learn, the changes are driven by forces largely beyond mankind's control. Climate changes are cyclic. We have lived through them before, as did the polar bears, and as will my grandchildren. A new climate cycle may have already started as I write this page.

Compared to the great heretics, Galileo Galilei and Martin Luther, my heresy is inconsequential. They acted with great courage against omnipotent religious and political authority to question seemingly unimpeachable orthodoxy, and in the process, unleashed our true intellectual potential. The movement they inspired became known as The Enlightenment, with the goal to "Dare to know! Have the courage to use your own understanding." (Immanuel Kant, 1784) I'm a minor heretic because I merely reject the omission of evidence, the alteration of evidence, and the complete fabrication of evidence to propagate the thesis that human-generated carbon dioxide emissions are causing the planet to dangerously and irreversibly warm.

This concern about the Earth heating up has become the defining issue of our times. It started in earnest about 35 years ago by environmentally conscientious activists. They warned that human-generated increases in the carbon dioxide (CO_2) content in our atmosphere would result in *global warming* with dire long-term consequences. The explanations given to support these claims were initially simplistic and often glaringly wrong, but they went mostly unchallenged except in academic circles.

Global temperatures were closely watched, and as they went up, the membership of the global warming movement grew by leaps and bounds. Momentum increased due to the efforts of celebrities, protestors, and new green political parties. Few of these people had any serious scientific expertise but they all demanded action. The feared crisis point of temperature increases, initially forecasted to happen sometime in the next 100 years, was revised down to only 50 years into the future.

To address this accelerated schedule, the world came together under a United Nations resolution in 1988 to form the Intergovernmental Panel on Climate Change (IPCC) with the mandate to investigate mankind's influence on global warming. By 1997, those efforts resulted in the Kyoto Protocol, a framework agreement to reduce human emissions of what are termed *greenhouse gases*, which are implicated in contributing to global warming. The greenhouse gas that was the primary target for reduced emissions was carbon dioxide. This put the global warming movement on a collision course with the global economy as every flame, even a candle, emits carbon dioxide and much of the global economy is dependent upon burning the fossil fuels coal, oil and natural gas.

Twenty years ago, the warming trend slowed significantly, despite carbon dioxide emissions growing faster than predicted. The reaction from the global warming movement was to rebrand with an emphasis on future global warming as *climate change*. The justification for the validity of future climate change was an erroneously claimed scientific worldwide "consensus" that global warming was mostly man-made and dangerous. Governments took action: trillions of public and private dollars were spent on green energy, green transportation, and green consumer products. The results of these programs were higher energy costs for the rich countries, higher food costs for the developing countries, and in places like London, UK, higher levels of unhealthy pollutants in the air. Fossil fuel consumption increased.

Forecasts of global temperature increases continued to fail. The movement shifted their focus, yet again, to a perceived but inaccurate increase in extreme weather-related events. Climate change was re-branded as *climate emergency*. Although human-emitted carbon dioxide represents only one of every 85,000 molecules in the atmosphere, those molecules were blamed for all weather-related emergencies, even though there was no statistical increase in the historical severity or frequency of these events. The climate emergency alarm was pulled because the activists believed that carbon dioxide emission levels were so high and growing so fast that we had little more than a decade before all hope for saving the planet from irreversible catastrophic temperature increases would be lost. This was propagandized to the point where the media presented climate change as an imminent existential world crisis. Then we taught this in our schools as fact and caused widespread anxiety among a generation who had not yet come of age.

From this, the 2015 Paris Agreement was born to drastically reduce greenhouse gas emissions, mainly carbon dioxide, which meant drastically reducing fossil fuel use. Rather than unifying the world around a common cause, the Agreement engendered chaos and disruption. Committing to the targets of the Paris Agreement affects which governments get elected, which countries are still strongly allied, and whether some countries remain united as countries. The poor populations riot over increased fuel prices, the oil producers are sued for meeting the demands of the oil consumers, and our youth accuse us of stealing their hopes for the future. Some major economies are addressing climate change, but the biggest carbon dioxide emitters by far have been incentivized by the Paris Agreement to maximize their greenhouse gas emissions as quickly as possible. The developing economies want a free pass on emission reductions and $100 billion per year in compensation from the developed economies that are expected to reduce emissions at their own economic peril.

Let's take a deep breath and try to figure this out. Before we undertake a jackhammering and restructuring of the entire world economy, we need to look at why we think we must do that. It seems the complete scientific justification offered by political leaders to the tax-paying public for fighting global warming, or climate change, or the climate emergency, is based on the findings and forecasts of the United Nations Intergovernmental Panel on Climate Change (IPCC).

It is incumbent on elected governments to explain to their citizens in simple terms the science behind climate change and humanity's role in it, and invite a fulsome diverse discussion and verification of the facts before committing their constituents to the full implementation of the 2015 Paris Agreement. That's how democracies handle trade agreements and military alliances. No government appears to be willing to do that for climate change, implying that either the science is too difficult to explain or that our politicians simply do not want to explain it because of the political liabilities. Governments with policy agendas wholly unrelated to climate change are often unable to meet their goals unless they are re-elected. If the electorate is concerned about climate change, governments must at least appear concerned as well. So our politicians defer to the IPCC. We deserve more justification than that.

Carbon dioxide emissions continue to grow. At the same time, the worldwide temperature databases stubbornly conflict with the climate emergency forecasts by refusing to warm by any meaningful amount for the last 20 years and counting. In the USA, where the best raw ground-based temperature data in the world exist, evidence suggests a cooling period may have already begun.

We need to know if carbon dioxide has already spent its ability to act as a greenhouse gas and whether peak global warming is already behind us. We need to know if greenhouse gas reduction initiatives will contribute to sending billions of people in the developing world back into the poverty from which they have recently emerged. We need to know if climate change is entirely natural, and whether all humans can do is to adapt to it as we have done since we stood erect. Most importantly, we need to know if the IPCC is a competent scientific organization and understand what is written in the 2015 Paris Agreement.

It is an understatement to say it would be prudent to work toward understanding climate change and the proposed remedies before our next step does more damage than good. Climate change is like a complicated engineering problem, and my primary objective is to explain it in a way that requires thoughtful engagement, but not post-secondary scientific training. My secondary objective is to explain historical climate change so you will have the tools to critically question the global warming information that we are bombarded with by politicians and the media. I hope this book better equips you to differentiate between someone's opinion and established science, and between faith and facts.

Climate change is complicated because there are many natural cycles and processes that influence our climate. On their own, each of these is relatively straightforward, but the complexity is in how they interact with each other, which is often uncertain. For this reason, the discussion often devolves into arguments over difficult mathematics (I promise you I will not introduce any equations), indecipherable graphs (I anticipate only clear illustrations), and endless hair-splitting debates of equipment accuracy, sampling methods and statistical analysis (I will stick to broad logical themes). The large, complex science of climate change will be broken down into small, simple concepts, and then the most important pieces will be put back together to solve the puzzle. I will focus on well-established science and accepted planetary history to derive what must be, and more importantly, what cannot be the driving forces of current climate change. I will also look at the IPCC and its many scientific failings, and the media's complicity with their misleading propaganda.

What you are going to learn by this method is that carbon dioxide cannot cause runaway global warming, and I shall use the long history of life on Earth to prove that. You will see that the varying level of the sun's energy generation is a primary driving force for climate change, backed by evidence from what is likely the longest-running scientific study in human history. I will also prove that modern Homo sapiens have lived and thrived in warmer times than now and experienced rapid temperature changes.

I will then pull back the veil of the IPCC to reveal far more politics than science. You will also learn that IPCC forecasts always fail because they have overestimated how sensitive climate change is to carbon dioxide, and that deep in their files they admit that current human additions of CO_2 have only a very minor contribution to increasing the greenhouse gas effect. I'll explain why the IPCC methods do not meet the standards of the Scientific Method and, therefore, do not qualify as scientific fact. You will discover that the famous hockey stick graph, which the IPCC used to claim that now is the hottest period in a thousand years, was investigated by the US government and found to be wrong, and that the IPCC attempted to hide evidence conflicting with the hockey stick. The real crisis for the climate emergency community is that global warming may have already reached a plateau, and the IPCC is having increasing difficulty hiding that.

You are going to learn about the major and costly initiatives your government may have funded to reduce greenhouse gases, and that they do not cause a reduction at all. In fact, these initiatives have the unintended consequences of further harming the environment, human health, and the poor. Your government's embracing of a non-existent climate emergency and the over simplified educational tools it uses may be why our youth are so despondent about the future. I'm going to include my own top 10 climate media stories you thought were true but are not, including the origin of the myth that 97% of scientists agree that global warming is mostly man-made and dangerous.

Then we will examine how the 2015 Paris Agreement is already past the point of total failure, and how it incentivizes China and India to quickly maximize their carbon dioxide emissions from poor quality coal. Fortunately, it does not matter from a climate change perspective, but these emissions are still a significant health hazard.

Lastly, I will suggest a way forward. The proposal is not uniquely mine, nor is it cheap or guaranteed to work. It does, however, have a 100% success rate for the entire history of humanity.

CHAPTER 1

Carbon Dioxide Deserves a Fair Trial

A climate change denier might simply state that global climate change is not happening, and a climate skeptic may say that global climate change is happening, but it is not as bad as you think. My heretical statement is that climate change is happening; it is not as bad as you think, and the reason is that we have been overestimating the role of one climate change driver while ignoring another. In this first chapter we will investigate the scientific reasons why the role of carbon dioxide as an agent of climate change has been overstated. In Chapter 2 we will examine the science of why the sun is a primary driver of climate change, and in Chapter 3 we will use anthropological history to tip the balance heavily in the sun's favour.

It might be shocking to you that there could be any question at all that carbon dioxide (CO2) is the villain behind climate change and the projections of catastrophic global temperature increases. It seems impossible that it could be otherwise, considering the United Nations-led Paris Agreement of 195 countries pledging to reduce carbon dioxide emissions to mitigate climate change. Surely on the long road from the Rio Earth Summit in 1992 to the Paris Agreement of 2015, somebody would have asked,

"Are you sure carbon dioxide can cause the temperature changes you are forecasting?" Many scientists and some environmentalists have answered, "No, carbon dioxide cannot do that." However, their voices have been drowned out by the media, politicians, and environmental groups, often with great vigour, lots of money, and personal attacks. There is also the much-cited "97% agreement among scientists," which we will invalidate in a later chapter that precludes many from even considering asking the question.

We will examine the basic principles of why global warmers claim carbon dioxide is the perpetrator and then look at why that is improbable, if not impossible. I will stick to my pledge to avoid the scientific quicksand of mind-numbing detail and battling PhDs. Instead, I will use basic science and logic. This discussion is intended to reveal what the media has hidden by their "dumb it down" approach. Somehow, journalists have decided that when it comes to matters of science, their average reader has a below-average capacity for comprehension. I refuse to accept that. I think the media underestimates the scientific capability and curiosity of their readers. The media doesn't even say "carbon dioxide" any more; they just say "carbon" as though they were the same thing. There are almost 10 million known carbon-based compounds.

The basis of the global warming argument is that the Earth stays warm in a similar fashion to a greenhouse, which uses a physical glass barrier to trap heat inside the building. They argue that the Earth's atmosphere contains specific gases, called greenhouse gases, which behave like the glass ceiling of a greenhouse. This trapping of heat by the greenhouse gases is termed the *greenhouse gas effect*, and the result is a warmer surface of the Earth. Carbon dioxide is a greenhouse gas, and the fear is that humans are putting too much of it in the atmosphere and that this excess will trap more heat and raise the temperature at the surface of the Earth catastrophically.

The analogy to a glass greenhouse is wrong, but the basis of the theory of the greenhouse gas effect is undeniable. However, in this chapter, we shall see the sensitivity of the Earth's temperature to increases in carbon dioxide has been greatly overstated. A greatly *understated* fact is that each future volume of carbon dioxide emitted has a significantly reduced effect on temperature than any previous emissions. This has been proven by the history of the Earth.

A further global warming argument is that high levels of CO2 in the atmosphere will dissolve in the oceans and acidify them, causing marine life to die off. I will explain how this is untrue. But first, a primer on the science of the greenhouse gas effect.

What is the Difference Between a Greenhouse and the Greenhouse Gas Effect?

Heat can be transferred by the following three different methods: conduction, convection, and radiation. Conduction occurs when heat is transferred from a hot object to a cold object when they come into contact with each other, like a frying pan on an electric stove element, or a spoon in a hot cup of tea. Convection occurs when heat is transferred in a current of fluid (it could be a liquid or a gas), such as a rising volume of hot air. A good example is to watch the formation of a towering thunderhead cloud (cumulonimbus). Radiation occurs when heat travels in electromagnetic waves and does not require any physical matter between the source and the object being heated. Radiation can travel through empty space or air, as sunlight does to warm the surface of the Earth.

The sun emits electromagnetic radiation in the form of ultraviolet and visible light. Electromagnetic radiation is part of our daily lives, but we tend to think of it in terms of individual applications, rather than a continuous spectrum of different wavelengths of energy. On the electromagnetic spectrum, the next just shorter wavelengths than the sun's radiation are X-rays (used in medical diagnosis), and shorter still are gamma rays (which if you detect inside a nuclear submarine means the nuclear power plant is leaking). The wavelengths that are just longer than sunlight are infrared (produced in your electric sauna to warm you up), followed by microwaves (used in cooking), and longer still are radio waves. All these applications use electromagnetic waves, but each application is a different wavelength. In this chapter we will classify electromagnetic radiation from the sun (ultraviolet and visible light) as short-wavelength radiation, and radiation given off by the Earth (infrared) as long-wavelength radiation.

Water also plays a significant role in the climate change debate. When it changes its physical state from ice at a temperature of 0°C to liquid water at a temperature of 0°C, it takes a lot of added energy, which is then stored in the liquid water. When liquid water evaporates into water vapour, the extra energy needed to change states is stored in the water vapour, and that energy is transported away when the water vapour is absorbed into the air. This is why sweating cools you down; evaporation is a cooling mechanism.

A greenhouse uses glass—but clear plastic also works—to allow the sun's short-wavelength radiation energy into the building, and to stop the loss of heat by trapping convection and water evaporation heat inside. The glass is incapable of stopping the Earth's long-wavelength radiation heat loss. The greenhouse gas effect isn't as simple. It is a very different process as it allows convection and evaporation heat to escape, and it stores only some of the long-wavelength radiation energy.

Of all the sun's radiation that impacts our planet, slightly less than half makes it to the surface and warms it up. A significant portion of the sun's energy is reflected to space by clouds, ice, and snow, and an approximately equal amount is absorbed by the atmosphere and then radiated back into space. The remainder, slightly less than half, of the energy from the sun that makes it to the surface of the Earth and does not get reflected to space, heats the planet up. In the diagram on page 14, we will track that energy as Sunlight in.

Once the surface is heated up by the sun, heat loss from the surface on a global basis is approximately 70% by a combination of the evaporation of water and the upward convection of that water vapour and warm air, and approximately 30% by long-wavelength radiation. Note, this involves a transition of this energy in terms of wavelengths. Short-wavelength radiation hits the surface of the Earth, the Earth absorbs this energy and heats up, then releases this energy in the form of long-wavelength radiation. Since air is a good insulator, virtually no energy is lost by conduction. On an energy balance basis, all the energy received from the sun to heat us up is also lost back to space to cool us down. This is what also happens on the moon, which receives the same amount of sunshine as the Earth, but the average temperature is minus 18°C. The average temperature of the surface of the Earth is about plus 15°C because of greenhouse gases.

Greenhouse gases are physically incapable of intercepting the incoming short-wavelength radiation from the sun. They can only absorb specific sizes of long-wavelength radiation from the Earth. Then they reradiate that energy in all directions, including back to the planet's surface where it again warms up the Earth. The absorption and release of this energy by the greenhouse gases are shown in the diagram below as *Back radiation*. When that energy heats up the surface of the Earth and is released again, it is shown as *Recycled back radiation*. This recycling of heat back to the surface of the Earth by the greenhouse gases, essentially using the same energy multiple times, keeps the surface of the Earth warmer than if there were no greenhouse gases. You will note from the diagram that the amount of energy getting recycled this way is about $2\frac{1}{2}$ times larger than the amount of direct energy from the sun.

This recycling is a very large and important cycle that is in balance. Global warming can only occur by the greenhouse gas effect if there is an increase in greenhouse gases, and at the same time, there is a surplus of long-wavelength radiation of the required specific wavelengths for the increased greenhouse gases to absorb. This greenhouse gas absorption limit due to the availability of long-wavelength radiation in turn limits the amount of back radiation, which limits the temperature of the Earth. We shall learn in this chapter how this absorption limit works, and we will use the history of the Earth to prove its existence.

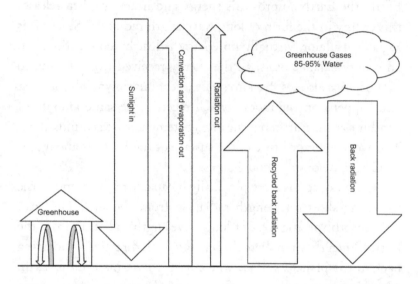

Figure #1: Comparison of a Greenhouse with the Greenhouse Gas Effect.

(1) Sunlight in: This is short-wavelength radiation from the sun that reaches the surface of the Earth. Greenhouse gases cannot absorb it.

(2) Convection and evaporation out: Approximately 70% of the short-wavelength energy received is used in evaporating water and creating convection currents, which transfer the energy to the upper atmosphere where it is released as long-wavelength radiation to outer space.

(3) *Radiation out: Approximately 30% of the short-wavelength energy received is converted to long-wavelength radiation and is released into space.*

(4) *Greenhouse: A greenhouse physically blocks the evaporated water and convection currents from escaping. There is no theoretical maximum to the heat that can be blocked.*

(5) *Greenhouse gases: The greenhouse gas effect is where the greenhouse gases absorb long-wavelength radiation of specific wavelengths from the Earth, and are back radiated to the Earth where they are reabsorbed. This recycling of long-wavelength energy is almost 2 $\frac{1}{2}$ times the short-wavelength energy received from the sun. Global warming can only occur if there is an increase in greenhouse gases combined with a surplus of long-wavelength radiation of the required specific wavelengths for the greenhouse gases to absorb.*

Here is a comparison of a greenhouse to the greenhouse gas effect in a nutshell:

- Both cause warming by interfering with the escape of heat to the upper atmosphere.
- Both warming results are limited by the amount of sunlight received at the surface of the Earth.
- A greenhouse traps heat from escaping and stores it inside the greenhouse; the greenhouse gas effect recycles heat from the ground to the atmosphere many times.
- A greenhouse has no theoretical maximum of the convection and evaporation energy it can trap within the greenhouse. 70% of the short-wavelength energy from sunlight is converted to convection and evaporation, all of which can be trapped in a greenhouse.
- Greenhouse gases have a limit on how much long-wavelength radiant energy they can absorb and recycle. 30% of the short-wavelength energy from sunlight

is converted to long-wavelength radiation, but only specific sizes of wavelengths can be absorbed by specific greenhouse gases.

- It only takes more sunlight to make a greenhouse hotter.
- It takes both additional greenhouse gases and additional long-wavelength radiation of the correct wavelength to warm up the surface of the Earth by the greenhouse gas effect.

Global warmers like the term "greenhouse gas effect" because they can shorten it to the "greenhouse effect." Their hope is that most of us mentally replace the physical glass ceiling of a greenhouse with an atmospheric gaseous ceiling made up of carbon dioxide, which traps more heat as more carbon dioxide is added, without limit. If you had no glass ceiling in a greenhouse, you would simply have, on average, a normal ambient temperature on Earth of approximately plus 15°C. But if you had no greenhouse gases, the temperature on Earth would be about the same as the moon—about minus 18°C. Therefore, we need some level of greenhouse gases to maintain a temperature for sustainable life on Earth.

If you have ever wondered how the Extinction Rebellion took root, please check out the NASA website for kids (climatekids. nasa.gov/greenhouse-effect/). The first illustration on the webpage is a depiction of the Earth inside a glass greenhouse. The second illustration is a glass greenhouse with plants growing on the inside and a snowman on the outside with the explanation: "Gases in the atmosphere, such as carbon dioxide, trap heat just like the glass roof of a greenhouse." NASA is not solely to blame for this wrongful analogy; our educational system taught this to my kids, and now, they are teaching the same to my grandkids. The takeaway message is: "The Boomers have put you into a hothouse with no limit to how hot it will get for their own economic benefit, and they don't care because they won't be around for the consequences."

Before these kids have learned how science works, some of them have given up on their education, their future, and have been exploited by the climate emergency crowd to lecture from the highest pulpits available to "accept the science." NASA and many others are complicit in teaching false science and creating fear in children.

I will attempt to convince you that additional carbon dioxide cannot cause the alarming global warming claimed by today's climate change conformism, that ocean acidification will not take place, and that carbon dioxide is not a pollutant. We are going to do this by imagining we have put CO_2 on trial. You can be on the jury, which is only fair as it is your tax dollars being wasted on "decarbonizing" and your children's futures being put at risk when there is so much reasonable doubt.

Global Climate Hysteria v Carbon Dioxide

The Arraignment

If human beings continue to burn fossil fuels, the resultant increase in atmospheric carbon dioxide—when it reaches the level of approximately 560 parts per million (ppm)—will cause the average temperature on Earth to increase about 3°C to 5°C above the average pre-industrial temperature. This will melt the polar ice caps causing sea level rises. The oceans will acidify, resulting in the loss of marine life. Carbon dioxide should be regarded as a pollutant and regulated as such.

Opening Statement by the Prosecution

Your Honour, members of the jury, the case against CO_2 is overwhelming. 195 independent nations, the United Nations Intergovernmental Panel on Climate Change, 97% of the world's climate scientists, and NASA, all tell us that man-made carbon dioxide released into the atmosphere will increase global temperatures significantly and dangerously. Nobel Peace Prizes have been awarded to some who have helped deliver this message. The increase in carbon dioxide that we have seen in the last 120 years could easily be replicated in the next 60 years even if carbon dioxide emissions are held constant, which they are not. They are still increasing. We have known for 120 years that we cannot keep releasing unlimited carbon dioxide into the atmosphere, and now that it's showing up in the oceans, it is time to stop.

I am only going to call one witness, and that is Science. What the jury will see is that the science is settled, there is no doubt of that.

We will begin with the first climate change model ever constructed, by Svante Arrhenius in 1896. A brilliant Nobel Laureate (1903 – Chemistry), he recognized that carbon dioxide was a greenhouse gas, and made calculations that a doubling of the amount of carbon dioxide in the atmosphere from the burning of coal would lead to a global temperature increase of 3°C to 5°C. As a Swedish citizen, he thought this would be a good thing for Nordic countries. However, at the 1896 rate of coal burning, this doubling of atmospheric carbon dioxide by human activities was some 500 years in the future. [1] At the publication of his paper, it is generally accepted that the atmospheric concentration of carbon dioxide was 280 parts per million (ppm); a doubling would be 560 parts per million.

Arrhenius was right in his calculations but wrong in his timeline. Temperatures will indeed increase by 3°C to 5°C, maybe more, with the doubling of carbon dioxide in the atmosphere to 560 parts per million. A simple Internet check will confirm than many pre-eminent climate scientists have checked his math and confirmed his calculations of temperature increase caused by the greenhouse gas carbon dioxide. As our scientific knowledge has advanced, these calculations have become more sophisticated, the

---

[1]   www.Brittania.com/biography/Savante-Arrhenius

measurements more exact, and the climate models much more rigorous.

The world has come together to form the United Nations Intergovernmental Panel on Climate Change (IPCC), under which the most talented minds concur with the basis of the Arrhenius climate model and the more advanced models that have updated his work. What Arrhenius got wrong was that this doubling would not take 500 years. Massive worldwide industrialization in the 20th century, notably after World War II, and the introduction of widespread use of liquid and gaseous hydrocarbons, such as crude oil and natural gas, along with the existing and expanding use of solid hydrocarbons, such as coal, means the doubling of atmospheric hydrocarbons is nigh.

120 years after the world's first climate change model, the atmospheric carbon dioxide content is 45% higher at about 410 parts per million. In 1897, mankind was releasing 1.6 billion metric tons of carbon dioxide into the atmosphere. In 2017, that number was 36 billion metric tons, over 20 times as much per year and still climbing significantly. [2] NASA both confirms the increase in atmospheric carbon dioxide and endorses the linkage between carbon dioxide increase and increased global temperatures. NASA also reports a global temperature increase of 0.8°C since 1880, with most of the warming occurring during the last 35 years. [3] Carbon dioxide atmospheric concentrations are currently increasing by 2.3 parts per million every year. [4] If this rate of increase is held constant, we will reach the Arrhenius doubling of CO2 in the atmosphere in 2085, and the most advanced computer modelling in the world forecasts that the global average temperatures will be about 4°C higher than in 1880.

---

2   www.Climate.gov/climate-change-atmospheric-carbon-dioxide
3   www.Climate.NASA.gov/evidence/global temperature rise
4   Climate.gov/climate-change-atmospheric-carbon-dioxide

The science is settled: carbon dioxide is a pollutant, and emissions of that pollutant by humans must be dramatically cut back, if not eliminated.

Defence Cross-Examines Science

Let's start with the last statement that carbon dioxide is a pollutant. Life on Earth can and has existed without sex, but it cannot exist without carbon dioxide. Bacteria, plants, and algae need carbon dioxide to start the food chain we all depend upon. Photosynthesis could not happen without carbon dioxide. In fact, commercial greenhouses add carbon dioxide to stimulate plant growth. Greenhouses can get 40% higher growth rates with CO2 concentrations between 1500 and 2000 parts per million. Since 1960, tree growth rates in Germany have increased substantially with higher levels of CO2 in the atmosphere. Even the IPCC forecasts faster-growing trees with more carbon dioxide in the air. Office buildings typically have two or three times the carbon dioxide level as the outside air due to human respiration, and we work inside those buildings most of our lives. Plants also do better in office buildings. Carbon dioxide is a colourless and odourless gas. The citizens wearing face mask filters in heavily polluted cities that you see in photographs are not wearing them because of carbon dioxide. They are protecting themselves from the smog.

Then let's look at the time gap between Arrhenius's paper of 1896 and 2020. Currently, we double our scientific knowledge every seven years. That's amazing. The Wright brothers didn't fly until 1903, and 69 years later, we had a man on the moon. Wouldn't you suspect that in 124 years we would have learned a lot more about the Earth's climate than Arrhenius knew?

Of course, we have, and I do not deny that the experts my Learned Friend has quoted know these things. For example, we know that of all the greenhouse gases, one gas is much more important than the others. Water vapour is a greenhouse gas and is responsible for 90% of the greenhouse gas effect. [5] Water vapour occurs naturally and cannot be controlled or regulated by mankind, which leads us to the inevitable conclusion that at least 90% of the greenhouse gas effect is beyond regulation or control. That alone should end this prosecution of carbon dioxide.

Prosecutor

Objection, Your Honour. This trial is about carbon dioxide, not water vapour. Besides, the charge against CO2 is not about the pre-industrial natural greenhouse gas effect, which was in equilibrium and provided us with a habitable planet. The charge is against only human-emitted CO2 that has upset that equilibrium and caused the planet to warm. That the greenhouse gas effect is 90% water is a red herring, Your Honour, because it is the anthropogenic CO2 that is triggering global warming, which is amplified by the 90% water vapour. If we did not add CO2 to the atmosphere, the existing water vapour greenhouse gas effect would not change.

Judge

Sustained! Defence, please stick to CO2. If you want to bring in water vapour, you will have to do it in another chapter.

Defence

My apologies, Your Honour. Let me rephrase and make the case of how small the contribution of carbon dioxide is to the overall greenhouse gas effect. The Earth absorbs part of the sun's short-wavelength radiation that hits our planet and emits 30% of all the energy that it absorbs back toward the atmosphere as long-wavelength radiation. Greenhouse gas in the atmosphere intercepts and absorbs some of this long-wavelength radiation. It then reradiates it in all directions, some of which is called *back radiation*—the portion of the energy that is directed back at the

---

5   Plimner, I: *Heaven and Earth* (Taylor Trade, Lanham, 2009) p 369.

surface of the Earth where it is reabsorbed. One expert calculation predicted that if the current carbon dioxide atmospheric content of about 400 parts per million was doubled, the amount of back radiation caused by carbon dioxide would go up only by 1 $\frac{1}{4}$%. This would cause a temperature increase due to only the increased carbon dioxide of about 0.3°C. [6]

[Loud murmurs from the courtroom.]

Judge

Order! You need to explain that! Are you saying that if we doubled the current CO2 from today's level, which is a further addition of 400 parts per million and would result in almost triple the pre-industrial level, the Earth's temperature would only go up by another 0.3°C? The temperature of the Earth has already gone up 0.8°C, with only an increase of only 120 parts per million CO2. I won't have you pulling a fast one in my courtroom, and I had better understand your explanation.

Defence

Your Honour, I am prepared to do that. As my Learned Friend indicated, when you include the amplification effect, which I shall explain shortly, of liquid water evaporation caused by the additional 400 parts per million carbon dioxide, the temperature increase is calculated to be 0.5°C—still not much. You see, the long-wavelength energy that the Earth radiates is in wavelengths between three and 100 micrometres, and most of it at less than 50 micrometres. One micrometer is one millionth of a metre, and an average human hair is approximately 100 micrometres thick. Carbon dioxide absorbs wavelengths of 14 to 16.5 micrometres, or much like the diameter of a strand of thin hair. Carbon dioxide captures only a narrow range of the Earth's thermal radiation, then sends some of which it captures back to Earth, and some

---

6   Plimner, I: *Heaven and Earth*, p 369-371.

is sent to the upper atmosphere and out to space. However, with each cycle of this long-wavelength energy, there is less and less back radiated energy in the 14 to 16.5 micrometre wavelength as the carbon dioxide radiates it in all directions, including back into space. This leakage of the 14 to 16.5 micrometre wavelength is what prevents a continuous buildup on the surface of the Earth or in the atmosphere of that specific wavelength. Eventually, you get to a place where extra carbon dioxide doesn't matter because the existing carbon dioxide captures almost all of the 14 to 16.5 wavelengths emanating from the surface of the Earth, and there is nothing for the additional CO2 to absorb.

Your Honour, I see you are a bit frustrated. Let's make an analogy. Near where I live, the Indigenous people catch salmon in fish weirs. These are vertical sticks placed strategically in the riverbeds in two rows. The salmon are trying to swim upstream, so they are going in only one direction. The first row of sticks is placed so that only a certain sized fish can get in, and the second row is a little bit upstream and closer together so that only a smaller sized fish can get out. So, fish bigger than a salmon cannot get past the first row (they swim around the weir), and fish smaller than a salmon can swim past both the first and second row of sticks. Salmon get caught between the two rows of sticks because they are small enough to get in, too big to get out, and do not want to turn around and go downstream. However, some of the salmon are smart or lucky; they turn around in the weir, swim downstream, get out of the weir the same way they entered it, and then swim around the fishing weir.

The first fisher builds his weir across one quarter of the river and catches a lot of salmon. It is a good spot. Two other fishers see this, and they each build another weir a little further upstream. Again, each covers a different quarter of the river. Both have a good catch but each about half of what the first fisher caught. Doubling the number of weirs resulted in catching the same amount of fish as the first weir. Four more fishers build four more weirs upstream, and each has a poor catch, only one quarter of what the first fisher caught. So, another doubling of the weirs again results in catching the same amount of fish as the first weir. What happened? Why did the second and third weirs each catch only half of the first and the next four catch only one quarter of the first? The number of weirs was multiplied by seven, but the fish caught was multiplied by three. Were the second through the seventh weirs defective?

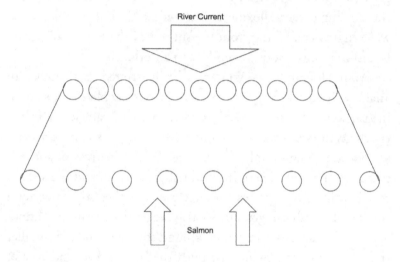

Figure #2: Fish Weir Diagram

Judge (enthusiastically)
    No. They just ran out of salmon! Sorry, I'm not supposed to comment.

Defence
    And that is why there is a limit to how much heat carbon dioxide can store in the atmosphere! As you add carbon dioxide to the atmosphere, the Earth eventually runs out of radiation between the 14 and 16.5-micrometre wavelength for the carbon dioxide to capture. When the amount of radiation of that wavelength generated by the Earth from direct sunlight is relatively constant

over time, as in Figure #1, the amount of back radiation downward by carbon dioxide is always less than the upward total of the radiation out plus the recycled back radiation. This is because the carbon dioxide only sends some of it back to the Earth, and the rest to all other directions. Each time the amount of carbon dioxide is increased, the increase of back radiation to the surface of the Earth is smaller. Eventually, if you increase the CO_2 high enough, there will be only an infinitesimal amount of 14 to 16.5-micrometre wavelength radiation left to capture.

Judge

Very good, Mr. Defence. Are you finished?

Defence

I'd like to go over that last point again, but with a little math involved. The fish weir story is an illustration of the difference between a linear and a logarithmic relationship. In a linear relationship, each time a fish weir is added, the same amount of additional fish is caught; seven fish weirs would catch seven times the fish as one weir. In a logarithmic relationship, each time you want to add the same sized catch, you need to double the number of weirs, and that doubling relationship never ends. If seven fish weirs catch only three times the amount of fish as the first weir, then it would then take 14 weirs to catch four times the fish as the first weir. It is well established in the scientific community, and acknowledged on the IPCC website, that the relationship between CO_2 and temperature is logarithmic, not linear.

I stated earlier that if the CO_2 concentration were doubled from 400 parts per million to 800 parts per million, the Earth's temperature would go up by 0.5°C, including the amplification effect of water vapour. Knowing how back radiation works logarithmically, how much more CO_2 would you need to increase the Earth's temperature by another 0.5°C? Any guesses?

Judge

I'll guess another 400 parts per million, bringing the CO2 concentration up to 1200 parts per million. I suspect you are about to tell me that is wrong.

Defence

Your guess is wrong, and it is a common error that the global warmers would like everyone to keep making. What you just described is a linear relationship, where every extra fishing weir would catch the same amount of fish as the very first weir. But the relationship between temperature gain on Earth and CO2 in the air is logarithmic. To increase the temperature gain by the same amount of 0.5°C, carbon dioxide would have to be doubled again from 800 parts per million to 1600 parts per million. This doubling of CO2 for the same 0.5°C temperature increase never stops. This ratio of a temperature increase with a doubling of CO2 is called the CO2 sensitivity. In this example calculation, the CO2 sensitivity is 0.5°C.

As I explained, the 0.5°C includes the amplification effect of water vapour, but there are two other water implications in all of this that could cause reduced temperatures. May I?

Judge

You may proceed.

Defence

If you doubled the existing approximately 400 parts per million more carbon dioxide in the atmosphere, it would cause an increase in back radiation. Most of that would fall on water and cause increased evaporation as liquid water covers approximately 70% of the surface of the Earth. Some of this increased water evaporation would act as a greenhouse gas, and this is called the amplification effect. Hence, the possible temperature increase above the 0.3°C from only carbon dioxide is caused by creating more water vapour to act independently as a greenhouse gas. The increased evaporation would also contribute to more cloud formation, which, as you know, cools the Earth by blocking sunlight. We don't know how much increased blocking of sunlight will occur because that science doesn't exist yet, and clouds are quite problematic for the IPCC climate models. The increased back radiation would also cause more convection currents from the surface of the Earth, which cools the surface by transferring heat to the upper atmosphere where it is radiated into space, but again, it is difficult to estimate by how much. So, increased greenhouse gas effect by additional CO_2 is amplified by evaporating more water vapour, which can act as a greenhouse gas itself, and is counteracted by that same water vapour forming clouds that will reflect sunlight or convection currents which will cause cooling of the surface of the Earth.

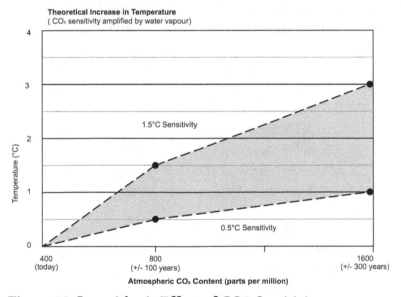

Figure #3: Logarithmic Effect of CO2 Sensitivity

Let's look at the lowest CO2 sensitivity presented by the Defence, which was based on a mathematical calculation. If the CO2 sensitivity was a 0.5°C increase in temperature for each doubling of the atmospheric CO2 concentration (lower curve), then when today's CO2 concentration of 400 ppm is doubled to 800 ppm, the temperature will have increased 0.5°C from today. If the CO2 concentration was increasing at 4ppm/year it would take 100 years to get to 800 ppm (the current rate of increase is much lower at about 2.3 ppm/year). To add a full 1°C from today's temperature, the CO2 concentration would have to be increased to 1600 ppm, which could take 300 years.

Now let's look at the highest CO2 sensitivity presented by the Defence, which is based on the observed temperature increase since 1880. If all the observed warming from 1880 to 2020 was due to increased atmospheric CO2 (meaning there were no global warming effects from any other factor) then the observed CO2 sensitivity would be 1.5°C for each doubling of atmospheric CO2 (upper curve). The global temperature would increase 1.5°C from today's temperature at 800 ppm and would increase to 3°C from today's temperature at 1600 ppm.

The above graph is not a climate change forecast model, it is a question: Why are the IPCC climate change forecast models predicting temperature increases much higher?

Prosecutor

28

Objection! Now he is talking about water vapour, computer simulations, and the IPCC! This chapter is only about carbon dioxide.

Judge
Sustained. Save those arguments for another chapter.

Prosecution Re-Crosses Science
We will not contest the fact that the relationship between CO_2 content and temperature increase is a logarithmic one that is accepted by both sides. What the Jury should recognize is that my Learned Friend is setting up an excuse to continue polluting the atmosphere with CO_2 on the basis that the next year's worth of CO_2 emissions will do less harm than the last year's worth of CO_2 emissions. Each year's worth of emissions does harm to the planet! We must stop the warming caused by CO_2 now when the temperature gains are worse than they might be in the future.

NASA satellites conclusively show that since 1970, the radiation from Earth to outer space in the carbon dioxide bandwidth of 14 to 16.5 micrometres have significantly diminished, and at the same time, the average temperature of the Earth has increased by most of that 0.8°C that has been previously acknowledged. Isn't that proof that carbon dioxide is absorbing that bandwidth of the Earth's radiation, then back radiating to Earth and causing global warming?

I must strenuously challenge the Defence claim that doubling the amount of CO_2 in the air will cause an increase in temperature of only 0.5°C. There is not an iota of observational data to support that ridiculously low number.

Defence Replies
I will concede to my Learned Friend that a CO_2 sensitivity of 0.5°C is at the low end of the range and that it is a mathematical estimate, not an observed estimate. However, I am using that

low sensitivity as a counterbalance to the high range of the 1896 mathematical estimate by Arrhenius, which is still in use today, and up to 5°C for a doubling of CO2. The Prosecution has already made the argument that since Arrhenius's time, the global temperature has gone up about 0.8°C, and the atmospheric CO2 content has gone up about 50%. The Defence is willing to concede that if no other factors were affecting the climate, and that CO2 alone was the cause of changes in the climate, the temperature increase of each doubling of CO2 would be approximately 1.5°C. The difference in CO2 sensitivity between the low-end mathematical estimate of 0.5°C and the observed estimate of approximately 1.5°C also demonstrates that there is room for other factors besides CO2 to affect global warming.

With respect to NASA, they do great data collection work, but if anything, the satellite data presented by the Prosecution confirms that any future global warming by further carbon dioxide emissions may be coming to an end. The fact that there is now much less of the 14 to 16.5-micrometre bandwidth reaching outer space means that the carbon dioxide concentration that exists today, about 400 parts per million, is already capturing most of the CO2 bandwidth that the Earth is radiating. It follows that more CO2 cannot back radiate much more energy because most of the supply of that bandwidth is already being absorbed. The satellites are like the fourth to seventh fishing weirs; the upstream weirs have caught most of the salmon, and there isn't much left to catch if you add more weirs.

There is another NASA satellite-based study [7] that measured how much long-wavelength downward radiation occurred from 1983 to 2007. The data showed that, except for a 1991 spike for the Mount Pinatubo eruption and the 1987 and 1998 El Niño spikes, there was minimal variation in downward long-wavelength

---

7   Stackhouse, P.W. et al. Feb 2011. "24.5-year *Surface Radiation Budget Data Set Released.* GEWX News Vol 21 No 1, page 10

radiation. In fact, the level in 2007 was the same as in 1983. The theory of CO_2 as a greenhouse gas states that increased CO_2 will cause increased downward long-wavelength radiation, and in a period where CO_2 increased substantially, there was no measurable increase in downward long-wavelength radiation. This suggests that the 14.5 to 16-micrometre wavelengths were mostly being absorbed and re-emitted already.

NASA's official positions on climate change appear to be a fulsome endorsement of the IPCC findings, but NASA does not conduct an independent body of work that confirms IPCC's findings. While NASA publicly supports the IPCC reports, some of the data they release undercut the IPCC climate theory.

My Learned Friend's concern is that I merely wish to kick the global warming issue down the road a few decades. My response is we may already be very close to the point, or perhaps at the point, where next year's CO_2 emissions will have negligible temperature effect. So, why are we trying to eliminate all CO_2 emissions by 2050? By the time we accomplish that, *if* we can accomplish that, increased CO_2 content in the atmosphere may be a non-issue. Our resources might be better spent addressing real issues as they appear, instead of very costly attempts to avoid issues that likely may never arise.

Defence Calls a Witness

Your Honour, my initial statements were to explain from a physical chemistry perspective why carbon dioxide cannot cause runaway global climate change. Now, I'd like to present three pieces of evidence that prove that carbon dioxide has *never* caused runaway global warming, yet in the past its concentrations were well above what the global warmers are saying are critical today. Then I will make a statement about ocean acidification by CO_2 and will explain why that is impossible, and even if possible, the effect on marine life will be nil.

My first and only witness that I am going to call is History.

Firstly, all carbon dioxide was created in the Earth's core and expelled by volcanoes when the Earth was very young, and the core was hotter. Then multi-cellular life on Earth began about 540 million years ago, and our best estimate of the CO_2 concentration at that time was about 5000 parts per million, and the temperature was 7°C warmer than it is today. The volcanoes had died down, life flourished, and CO_2 was used up by evolving plant life.

440 million years ago, the global temperature abruptly spiked to 1°C below today's temperature (about an 8°C drop), and then abruptly rebounded. In geological terms, 40 million years is abrupt. There was no corresponding change in atmospheric CO_2, which was about 12 times higher than today. The global temperature changed significantly both up and down, while CO_2 remained constant at a high concentration.

Beginning about 360 million years ago, massive plant growth occurred that resulted in present-day coal deposits. This photosynthesis consumed the atmospheric CO_2, and by 290 million years before the present day, the CO_2 level was drawn down to 400 parts per million, about the same as today. Plants grow best, as modern-day greenhouse growers know, at CO_2 levels between 1500 and 2000 parts per million. At the same time as the CO_2 dropped from 5000 parts per million to 400 parts per million, the temperature dropped from 6°C above today's temperature to 1°C below, and a major glaciation occurred. Global cooling had occurred in tandem with a significant drop in atmospheric CO_2 to low concentrations.

The glaciation was over by 250 million years ago, the world warmed up to about 8°C hotter than today, but the CO_2 was still at about 400 parts per million. Global warming had occurred without any change in atmospheric CO_2 at low concentrations.

However, by 140 million years ago, the CO_2 level had increased again to 2500 parts per million, six times higher than today. It is suspected that mushrooms had by then evolved enzymes that could decompose lignin in wood and release CO_2 back into the atmosphere. While the CO2 level was rebounding, the Earth's temperature dropped from 8°C hotter to only about 2°C warmer than today. Global cooling occurred, while CO2 levels increased to significantly high levels.

During the last 140 million years, the atmospheric CO2 levels continued to decline to their lowest levels ever–180 parts per million–just 18,000 years ago. This level was dangerously close to the plant CO2 starvation level of 150 parts per million, below which all plant life would die off. Global temperatures simultaneously rose to 9°C hotter than today, then cooled down to be 1°C lower than today, and another ice age occurred. In this period where carbon dioxide decreased to dangerously low levels, the Earth experienced both global warming and global cooling to both extremes.

In all this time, 540 million years, with extreme changes in both the atmospheric CO2 levels and the temperature of the Earth, there was no predictable correlation between warmer temperatures and higher CO2 levels. Sometimes they moved in tandem in the same direction, sometimes they moved in different directions, and sometimes they moved at random. [8]

The Earth's orbital cycles most likely played a role in all these climate changes, probably plate tectonics as well, but carbon dioxide was not the driver of past climate change. I will readily admit that the margin for error in CO2 and temperature measurements increases with the length of time looking back, but these estimates are largely accepted. Nevertheless, it is the magnitude and relative directions of the changes that are important in this evidence, not the accuracy of the estimates. Trends of CO2 concentrations in the atmosphere and the temperature on the surface of the Earth have a half-billion-year history of moving independent of each

8   Moore, P. 2015: *The Positive Impact of Human CO2 Emissions on the Survival of Life on Earth.*

other. There is no evidence that CO_2 can cause runaway global warming. Furthermore, there *was* no runaway global warming.

Carbon dioxide cannot cause runaway global warming because the Earth, unique among planets, has liquid water. As greenhouse gases, including water vapour, increase back radiation to the Earth, the free water on Earth absorbs this heat as latent energy and vaporizes. Convection currents then send this energy to the upper atmosphere where it is radiated into space. Evaporation cools the surface of the Earth.

My first piece of historical evidence clearly shows there is no link in geologic time between major climate shifts and carbon dioxide in the atmosphere.

The second piece of historical evidence is that the Earth has never experienced a runaway greenhouse gas event with many times more CO_2 in the atmosphere than exists at present.

For my third piece of historical evidence, we will look at CO_2 atmospheric concentrations from where we left off about 18,000 years ago, which was 180 parts per million. The conveniently named Last Glacial Period began to end 18,000 years ago, and the Earth warmed up until today's temperatures were reached about 11,700 years ago. At that time, the atmospheric CO_2 concentration was 260 parts per million.

Interestingly, the data show that during the 6,300 year interval above there was a mostly smooth, gradual temperature increase of 8°C, and then with a delay of several centuries it was followed by mostly smooth, gradual CO_2 increase of 80 parts per million. That is to say, the Earth heated up first, then lagging many hundreds of years later the CO_2 levels in the atmosphere rose. My third historical piece of evidence is this: when recovering from the Last Glacial Period, an 80 parts per million CO_2 increase did not cause a global warming of 8°C. Instead, it was the other way around where a global warming trend of 8°C caused an increase of 80 parts per million CO_2 concentration in the atmosphere, which ultimately reached 280 parts per million prior to industrialization. [9] This is the reverse order of what the global warming community is predicting should be happening now.

Your Honour, there is a perfectly good explanation for this. Cold water can absorb more gas in solution than can warm water. That's why today, there are good fish feeding grounds at both the poles, the cold water holds more CO_2 for plankton photosynthesis and more oxygen for the fish. The oceans of the world contain 50 times more CO_2 in solution than the atmosphere. When the last ice age began to end, the warming water gradually released some of the CO_2 it had absorbed during the Last Glacial Period. Simple.

My last statement is about the effect atmospheric CO_2 has on the oceans. The global warmers claim that as the oceans absorb more CO_2 from the atmosphere, the mixture of CO_2 and water will form carbonic acid, and the oceans will become more acidic. While this has technical merit in freshwater lakes, it has little effect on the oceans because they have a built-in resistance to acidification due to salts. The strength of an acid is measured on a scale called pH, where a pH of 7 is neutral between an acid and a base. The range from 0 to 7 is acidic (or example the acid in your car battery); and the range from 7 to 14 is basic (for example bleach or many other household cleaning products). A hot tub would have a water pH level of about 8.

The amount of acid that would cause a change of 1 pH in fresh water of 7 pH, would cause only a 0.003 pH change in sea water of 7 pH. Chemists call this resistance to pH change *buffering*. It would take 330 times the acid to get the same pH change in neutral sea water as it does in neutral fresh water due to the salt buffering of the sea water. Also, a widely accepted rule is that when the atmospheric CO_2 concentration doubles, the amount absorbed in the oceans increases by only 10%. Not only is the pH of the oceans almost unaffected by increases in atmospheric CO_2

---

9   Ibid.

content, but the natural pH of the oceans varies greatly around the world due to currents that cool at the poles and pick up CO_2, and then degas in warmer climate zones.

The ocean acidification alarmists claim the pH of the oceans has already dropped 0.1 pH units, from 8.2 to 8.1, and this will increase to a drop of 0.3 pH units by 2100. What isn't released by the same groups is that the natural pH of the oceans varies geographically from 7.5 to 8.4, and the worldwide average is about 8.1. How did they calculate a worldwide drop of 0.1 pH units when the natural variation is already 0.9 pH units? How can a drop of 0.3 pH units cause an unprecedented ocean extinction event when the normal range is 0.9 pH units?

And the marine life, Your Honour? Shellfish evolved when the CO_2 in the atmosphere was 12 times higher than now; we have had clams for 500 million years. Lab testing indicates shellfish may even be better off with more CO_2, as that is where the shell-building material comes from. [10]

The Defence rests.

Prosecution Closing Argument

Let's remember a few facts that my Learned Friend has introduced. It is apparent that since the industrial revolution started, the increase in atmospheric CO_2 has risen about 75 times faster than the previous natural CO_2 degassing from the warming oceans after the Last Glacial Period. Clearly, human activity has contributed more to the CO_2 emissions than recent ocean degassing. In the last century, we saw an increase in temperature on Earth of about 0.8°C, which is about six times faster than the temperature rise after the Last Glacial Period. Since the start of the industrial revolution, there is a correlation between a 75 times faster atmospheric CO_2 increase and a six times faster temperature increase. The Defence would like you to ignore that mankind has

---

10   Moore, P. 2015: Ocean "Acidification" Alarmism in Perspective

contributed to a global increase in carbon dioxide, and the result has been an accelerated increase in global warming. There is a name for people who ignore the facts; they are called Deniers.

What about the increase in hurricanes, forest fires, floods, tornadoes, polar vortexes, and especially heatwaves? What about rising ocean levels, melting ice caps, starving polar bears, and the spreading of malaria to new areas? All these things have gone up as the burning of fossil fuels has gone up; everyone knows that. Your duty as the Jury is to stop the burning of carbon-based fuels so that your kids and grandkids can have a planet to inhabit.

Defence Closing Argument

I would like the members of the Jury to reject the Prosecution's projections of fear based on unsupported claims and instead embrace the arguments based on scientific logic. I think we all realize now that carbon dioxide is essential to all life on Earth, and that we routinely exist in it indoors at many times its current outdoor level. Without CO_2 acting as both a greenhouse gas and as a necessary agent of photosynthesis, the Earth would be either frozen or barren of life, or probably both. CO_2 can't be considered a pollutant as it is essential to all life; otherwise, we would have to do the same for fresh air and fresh water.

Likewise, the concept of increased carbon dioxide in the atmosphere causing the oceans to become more acidic and harming marine life is a non-starter. The oceans are heavily buffered, so they are resistant to pH changes. The rate of absorption of CO_2 from the air into sea water is very low, and marine life both evolved and currently live in naturally occurring variations of pH in the oceans that are greater than the range that global warmers claim will cause their extinction.

In the geologic record of 540 million years, there is no correlation between CO_2 content in the atmosphere and significant temperature changes; CO_2 concentrations have been both high and low during much warmer periods. In 18,000-year ice core records, we can see that as cold water warms up, it degases CO_2, confirming that global warming causes increased atmospheric CO_2 content. This is the reverse of the claim that increased CO_2 causes global warming. The 18,000-year ice core records clearly show that the rate of temperature increase in the century is not unique and that there have been many greater spikes of temperature change.

Carbon dioxide is a greenhouse gas, and up to a certain point, increased concentrations do help warm the Earth's surface. However, that point is reached when all the infrared radiation between 14 and 16.5 micrometres emitted by the Earth is absorbed. The CO_2 level at the end of the Last Glacial Period was so low that any additional CO_2 would act as a greenhouse gas as there was most likely more infrared radiation of the correct band being radiated from the Earth than there was CO_2 in the atmosphere to absorb it. Satellite data indicate that today, we are very near the point where increased CO_2 in the atmosphere may not have a significant further greenhouse effect. We also must remember that the relationship between CO_2 and temperature is not linear; it is logarithmic. The CO_2 concentration has to be doubled again each time for the same temperature increase.

Let's think about what that means. If from pre-industrial times, the CO_2 level in the atmosphere has gone up by 120 parts per million, and let's say the CO_2 increase and nothing else have caused the entire temperature increase of 0.8°C, then it will take an additional 240 parts per million, for a total of 640 parts per million, to achieve about a total 1.6°C temperature rise since pre-industrial times. I'll repeat this is neglecting all the non-CO_2 factors that may have contributed to the 0.8°C temperature rise we have already seen, and that the CO_2 sensitivity rule results in an additional 240 parts per million would add another 0.8°C. At the current rate of CO_2 increase in the atmosphere, this would take over 100 years.

However, there is no scientific proof that CO_2 caused the temperature increase; correlation does not prove causation. It is difficult to differentiate between CO_2 greenhouse gas temperature increases and water vapour greenhouse gas temperature increases, the latter of which makes up 90% of the greenhouse gas effect. Between the minimum CO_2 doubling sensitivity based on calculations of 0.5°C and the maximum based on observations of 1.5°C, there is much room for other factors to cause global warming. There is more than reasonable doubt that human-emitted CO_2 is responsible for the 0.8°C temperature increase in the last century, and it's unknown how much of that temperature increase was from natural changes.

I also want to state that while I learned from the local Indigenous people how fish weirs work, I would like to clarify that they would never overfish. In fact, the weirs they showed me were to catch salmon, count them, look for abnormalities, and then release them. Their weirs are for conservation purposes. I borrowed them for illustration only.

And I am a heretic, not a denier.

Judge's Instruction to the Jury

This trial is only about whether human-sourced carbon dioxide is guilty of causing current climate change and if it is likely to cause significant future climate change. It is not about the consequences of climate change, and it is not about the sources of human-emitted carbon dioxide. It is not about the opinions of other parties on the charges against CO_2. It is about the known facts regarding CO_2. You will render a verdict on the following:

1) Has human-emitted CO_2 materially contributed to global warming since pre-industrial times until today?

2) Will further human increases in the CO_2 content in the atmosphere cause material increases in the Earth's temperature in the future?

3) Will CO2 acidify the oceans?
4) Is CO2 a pollutant?

Based on your verdicts, you, the taxpayer, will have to either continue to spend hundreds of billions of dollars immediately to fight climate change caused by human CO2 emissions, or you may be required to fund some unknown amount of money over the next century to adapt to the consequences of naturally occurring climate change as it manifests itself. Your verdict is not to be based on any other perils of any pollution unless you consider that pollution to be carbon dioxide.

The Jury should now retire to deliberate.

CHAPTER 2

The Sun is the Smoking Gun of Climate Change

Imagine you are on a camping trip. You are outside in the pre-dawn greyness, and you feel chilled. You have started a campfire to keep warm, but the chill won't go away. Then the sun fully climbs above the horizon, and you face east, close your eyes, and stop moving for a few minutes. The sun's rays warm you, and you feel good.

The next morning, you have the same routine, and you anxiously await the sunrise. The darkness is chased away, and you face east, but you are stilled chilled. There is no direct sunshine on your face because today there are low, dense clouds. You get closer to the campfire, which helps, but you also decide to put some more wood on the fire. Soon you are warm and comfortable in your camp chair, despite the gloomy clouds.

It is safe to say that within this 24-hour period, the energy output from the sun has not changed, and the distance that energy has travelled from the sun to the Earth is the same. Less of the sun's energy fell on your face on the second morning, and you felt chilled because the clouds reflected much of the sun's energy into space. This is not climate change, only a difference in the weather, but it illustrates an important concept: the sun's output can be constant while the amount of energy that reaches the surface of the Earth can materially change, just because of cloud conditions. Moving closer to the campfire increases the warmth we feel from it dramatically without changing the energy output of the fire, and increasing the size of the fire helps you warm up too.

Now, imagine the time frame isn't 24 hours. Instead, you and your ancestors have lived near your campsite for many centuries. Over that time, they have observed that the average temperature can go up or down by a few degrees centigrade, and this affects both the crops they grow and their culture. Your scientists observe that sunspots are an indicator of how active the sun is, and that they increase in frequency with the energy output of the sun that is directly heating the Earth. They figure out that sunspots are also an indicator of the strength of the sun's magnetic field and learn that this changes the cloud cover of the Earth. During the day, clouds can reflect from 10% to 90% of the sun's energy (high thin clouds vs. low thick clouds, respectively) [11] into space, whereas at night, they can inhibit heat from escaping the surface of the Earth. Because of this, you and your ancestors discover that cloud type and cover is the Earth's natural thermostat.

You now understand the basics of shorter-term natural climate change. The rest is just solar physics and thermodynamics.

The last part of our thought experiment is to imagine the time frame is much more than many centuries but extends back in time hundreds of thousands of years. The shorter-term solar sunspots still occur, but now much longer-term variations affect our climate. These are the cycles of the Earth's orbital patterns, which comprise some years wherein our orbit is closer to the sun, some years wherein it is further away, and the changing tilt of the Earth, which gives us summer and winter seasons. Our planet also wobbles, which makes the climate puzzle a little more interesting. You still live in the same spot on Earth, but now, in the coldest case

---

[11]   www.sjsu.edu/faculty/Watkins/cloudiness

scenario. The sun's energy output is at the cyclical lowest, causing the Earth's cloud cover to be at its cyclical heaviest, and at the same time, the Earth's annual orbit is furthest from the sun. The Earth has tilted on its axis to where the winters are not as severe but still freezing, while summers are cooler and have fewer frost-free days. At your campsite it is already very cold, life is hard, and the Earth may already be in an ice age. But then, one of the pinwheel arms of our Milky Way galaxy rotates between us and the sun, sending a haze of cosmic dust and increased cosmic rays that further impede the sun's energy from reaching the Earth.

Now you understand the current thinking on how naturally occurring long-term climate change could become a major ice age.

The IPCC largely dismisses variable solar activity in their climate change forecasts [12, 13] as having little or no impact on changes in the Earth's temperature because the variations are deemed to be minor, and the orbital variations are much too long for their forecasting time frame. A more candid answer might be that they cannot yet model the effect of the sun's magnetic field on cloud formation, so it would merely highlight a significant shortcoming of their computer forecasts. Even if they could do that analysis, the results of all computer forecasts are not scientific evidence and do not constitute observations of climate change. I'll expand on that in a later chapter.

Instead, we will have a more detailed look at the basic science described above of how solar energy variance that ultimately hits the surface of the Earth affects the global climate, and why these effects are materially important. The aim of this discussion is to refute the claim that solar system variations are inconsequential to climate change. They are not inconsequential; they are vastly more significant than the exaggerated claims of the future carbon

---

[12] Plimner, I: *Heaven and Earth,* p 101, 112, 115, 121, 132.

[13] IPCC. *Assessment Report 5 Climate Change 2013. The Physical Science Basis,* Ch. 9 "Evaluation of Climate Models".

dioxide induced temperature increases from the global warming community. As promised, I'll avoid the hair-splitting of data and accuracy of measurements. The purpose here is to describe the driving mechanisms of climate change, and we will see proof of their work in a later chapter. We will also review the Earth's orbital cycles and galactic rotation so that we can properly separate longer-term climate changes such as major ice ages from shorter-term climate changes such that are the subject of the current global warming debate.

The Sun is the Gun; Sunspots Are the Smoke

The sun provides almost all the energy to the biosphere on Earth. There is a very small component of heat from geothermal energy that is very localized where it is noticeable, for example, terrestrial volcanic activity in Antarctica or Iceland. Geothermal energy is generally very small throughout the rest of the world; however, we don't know much about subsea or sub-ice volcanic activity. It is the sun that drives photosynthesis, desalinates ocean water into rain, and powers the atmospheric and oceanic currents. Too much of the sun's energy and the blue from the blue planet would be boiled away, while too little of the sun's energy and the blue freezes solid. That narrow temperature range of liquid water needed for life is reason enough to consider even the smallest changes in solar energy reaching the planet's surface as significant.

Our sun is made up of dense gas, mostly hydrogen (about 70%) and helium (about 28%). It started as solely hydrogen, as this is the only element provided to the universe by the Big Bang. The basic theory of the Big Bang is that, "In the beginning, there was nothing, which exploded." (Terry Pratchett, 1992) Gravitational attraction caused some hydrogen to group together, a process which was accelerated as the ball of gas grew. Eventually, the ball of hydrogen got so big that at the centre, the compression forces created a temperature of 15 million °C, which is ideal for nuclear fusion to occur.

At this temperature, the hydrogen is in the fourth state of matter, which is called *plasma* (the other three states being solid, liquid, and gas). Plasmas also exist on Earth within lightning bolts and in the spark generated in the spark plug of your car. Nuclear fusion is when two atoms are combined and energy is released, as opposed to nuclear fission where an atom is split, and energy is released. The core of the sun, approximately 25% of its total diameter, is a fusion reactor, which takes two hydrogen atoms and makes a helium atom and releases a lot of energy. The energy is mostly in the form of electromagnetic radiation (photons or light), and a subatomic particle called a neutrino (we don't need to follow up on them). Between the 25% diameter and the 95% diameter is a zone where the photons move slowly upward by radiation toward the outer layer of the sun. The outer layer is the final 5% of the diameter of the sun, but that is still a thickness of 200,000 km. This is almost 16 times thicker than the diameter of the Earth. In this layer, the light energy makes its way to the surface by convection currents in the hot, dense gas, and then is released into space. The surface of the sun is a much cooler 5,500°C.

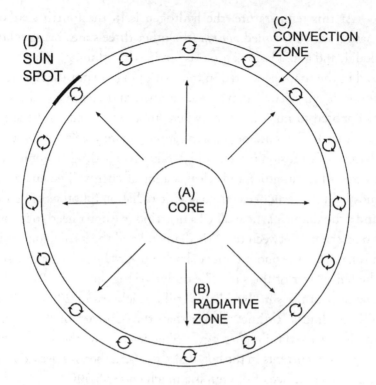

Figure #4: Cross Section of the Sun

A. *The Core of the sun is a nuclear fusion reactor where hydrogen atoms combine to form helium, and energy is released.*

B. *The Radiative zone is where the energy in the form of photons slowly work their way outwards.*

C. *The Convection Zone is where the photons are moved to the surface of the sun in convection currents of gases, where they are released into space.*

D. *When the Convention zone speeds up the release of photons, sunspots appear as dark blemishes on the surface.*

If the convection currents in the outer layer are sped up, an increase in light energy is released at the surface of the sun. The speeding up of the convection currents appears on the surface as increased sunspots, which look like small, dark blemishes. They are darker because sunspots are cooler than the surrounding sun surface. Sunspots are like a speedometer on the sun's convection currents in its outer layer. A faster convection current means more electromagnetic photons (light) energy released into space, and the sun is characterized as being more active.

A highly active sun can put out 0.1% more energy than an average sun, and a very quiet sun can put out 0.1% less energy than average. The difference from solar maximum (active) to solar minimum (quiet) output is 0.2%, and because the sun puts out so much energy, this small difference of 0.2% has been calculated to make a 0.45°C difference on the Earth's surface. [14] That small amount of natural temperature change is significant in the context of contemporary climate change, yet global warmers ignore it because they are predicting up to 10 times that temperature increase caused by increased carbon dioxide.

High activity of sunspots indicates the sun is more active and releasing more light energy, but that is not the biggest effect of a more active sun on climate. The sun's large magnetic field affects the Earth. We are constantly bombarded on Earth by cosmic rays of protons that have been sent our way by supernovae (exploding stars). When the sun's activity is high, which is indicated by higher sunspot number and size, the sun's magnetic field is also stronger. This magnetic field deflects some of the cosmic rays from us. A 2017 experiment at the CERN facility [15], the largest particle physics laboratory in the world, finally confirmed a long-held hypothesis that cosmic rays that hit our atmosphere promote low-altitude clouds. Reduced sunspot activity is a signal that the sun's magnetic field is weaker, which allows more cosmic rays into our atmosphere, and that causes more low-altitude clouds to form. As

---

[14] Plimner, I: *Heaven and Earth,* p 122.

[15] Svensmark, H. et al. 2017: *Increased ionization supports growth of aerosols into cloud condensation nuclei.*

we saw from our camping thought experiment, low-level clouds block some of the sun's light energy, and the Earth cools. The opposite is also true; more sunspots would result in fewer low-altitude clouds and more warming on the Earth's surface.

It has not yet been established how many more clouds and how big a cooling/warming effect can be caused by a change in the sun's magnetic field, but it must be larger than the direct impact of the sun's energy output, which, as noted above, has been calculated as a 0.45°C change from solar maximum to solar minimum. In our thought experiment, the camper probably would not feel a 0.45°C difference during the day but would immediately feel cooler when the sun disappeared behind a cloud. Generally, clouds reflect up to 60% of the sun's energy back into space, so even a small change in the Earth's cloud cover can make a huge difference in the Earth's temperature. Clouds matter a great deal in the climate change debate, and the IPCC admits that, "Confidence in the representation of clouds and aerosols remains low." [16]

Here's how the cosmic rays to clouds linkage works:

- Cosmic rays, mostly high-energy protons, strike the atmosphere and remove electrons from gas molecules and aerosols (small solid or liquid particles suspended in the air). This produces ions, that is, positive and negative charged molecules in the atmosphere.
- The ions help clusters of mainly water molecules and sulphuric acid to form and become stable. This is called nucleation. The small clusters then grow by a factor of many million times, which other ions created by the cosmic rays also help to accelerate.
- When the clusters get big enough, they become sites where liquid water droplets form clouds.

---

[16]   IPCC. *Assessment Report 5 Climate Change 2014 Synthesis Report,* p 56.

We now have the chain of fewer sunspots, less magnetic shielding from the sun, and more cosmic rays hitting our atmosphere. This creates more low-altitude clouds that can reflect about 60% of the sun's energy from reaching the Earth's surface. This is superimposed on the simultaneous linkage of fewer sunspots, and a less active sun, resulting in a lower intensity of light energy being emitted by the sun, and up to a 0.45°C temperature decrease on Earth. Since the sun is the only source of heat for most of the planet, sunspots are something we should not ignore.

While the IPCC climate change models ignore sunspots, many scientists have been tracking sunspots in what is probably the longest-running scientific experiment in human history. After Galileo invented the telescope, he spotted his first sunspot in 1607. Since the early 1700s, we have continuous unbiased sunspot data recorded over 10 generations. A predictable, repeatable 11-year cycle has been observed for the last 300 years, and the amplitude of those cycles varies.

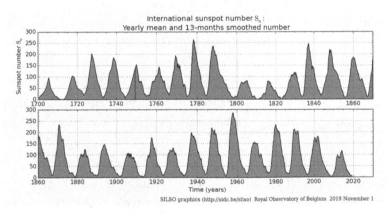

Figure #5: Sunspot Histogram Since 1700

What is interesting to do is to look at the graph of sunspot activity and relate it to known historical climatic events to see whether there is a correlation:

- From 1779 to 1818, a world-famous astronomer, William Herschel, noticed a correlation between sunspots and the price of wheat in England, as independently published by Adam Smith in the *Wealth of Nations*. Years with more sunspots produced better harvests, and the wheat price dropped.

- In 1799, what could be the world's first ice hockey game was played in London when the Thames froze over.

- The War of 1812 between Canada and the United States (1812 to 1815) was fought in bitterly cold weather.

- The decades from the 1940s to the 1970s are often referred to as proof that rising Earth temperatures are tied to rapid industrial development and increasing CO_2 emissions. Sunspot activity soared in this period.

- The high Arctic was warmer in the 1940s than it is today. [17]

- Global Warming was rebranded as Climate Change early in the 21st century because the temperature increase paused for 20 years as sunspot activity declined back to the level of the 1910s.

These anecdotes don't prove that current global climate change is directly related to sunspots, but they help point out that completely dismissing sunspot activity from the debate could be an error.

The easily observable 11-year solar cycle masks more complicated solar dynamics. There also appears, subject to much current debate, that there is an 88-year cycle, a 200-year cycle, and a 2,400-year cycle, which independently affect the level of the sun's activity. When some of these cycles combine to significantly *increase* the average solar activity for many decades, it is called a

---

[17] Yamanouchi, T. 2010: *Early 20th Century Warming in the Arctic: A review*.

Grand Solar Maximum. We recently just ended a Grand Solar Maximum that started in about 1914. On the other hand, when these solar cycles combine to significantly *decrease* the average solar activity for many decades, it is called a Grand Solar Minimum. The complexity of the cycles, and a lack of understanding of what causes them, makes predicting Grand Solar Maximums and Minimums difficult. NASA predicts we are entering the weakest 11-year solar cycle in 200 years in 2020 [18], and one pre-eminent geophysicist is predicting the start of a Grand Solar Minimum in 2020. [19]

We are now going to review some more solar-galactic systems science so that we can address why the ice ages happened.

The Earth is a Moving Target: Milankovitch Cycles and the Milky Way Rotation

We have it in our mind's eye that our solar system has the sun fixed at the centre with the planets unfailingly orbiting in prescribed paths, each complete circuit identical to the millions or billions that preceded it. This is an oversimplification as the planets' orbits vary slightly and can affect the climate on each planet. We are going to discuss the Earth's orbital changes and their effects on climate, including their probability as a trigger on the initiation and secession of ice ages. While this is still interesting climate science, each of the cycles is too slow to have a material bearing on 20th and 21st-century climate change.

The Earth has a slightly elliptical orbit that, over a period of about 50,000 years, becomes a bit more elliptical, and after 100,000 years, the cycle is back to an almost circular orbit. This pattern of change in the orbit is due to the gravitational pull of Jupiter and Saturn. The difference appears undetectable if drawn on a regular piece of paper, but in the scale of our galaxy, the

---

[18]   www.Nasa.gov. 2019: Solar Activity Forecast.

[19]   Zharkova, V.V. et al. 2019. *Oscillation of the baseline of solar magnetic field and solar irradiance on a millennial timescale.*

distances matter. Currently, the Earth's orbit is near its minimum elliptical shape and the difference in the total energy received by the sun within one year at our closest point of orbit compared to the farthest point of orbit from the sun is about 3%. When our orbit gets to the maximum elliptical shape in about another 50,000 years, the sun's energy hitting the Earth will be about 30% more intense when we are closest to the sun in our orbit than when we are furthest away. When Milutin Milankovitch developed these calculations in the 1920s, many agreed with him that this changing ellipse of the Earth's orbit could trigger an ice age. Today, we have established a pattern of glaciations over the last 800,000 years and an almost direct correlation of Antarctic temperatures from ice cores for the last 300,000 years with this 100,000-year Milankovitch cycle. We have had major glaciations approximately every 100,000 years. [20]

The second Milankovitch cycle is somewhat faster-paced. Currently, the Earth's axis of rotation through the poles is tilted about 23 $1/_2$ degrees from the flat plane of our orbit around the sun. However, over 21,000 years, it can vary from a minimum tilt of about 22 degrees to a maximum of 24 $1/_2$ degrees, again due to the influence of nearby planets. The change in tilt alters the amount of solar energy received at the poles compared to the equator. The tilt of the planet's axis is the cause of our seasons, and the more tilted the axis is, the more pronounced the seasons become. At a higher tilt, summers are hotter, and winters are colder, and at a lesser tilt, summer and winter both have more moderate temperatures. These cooler summers may have been another factor in the buildup of ice sheets preceding an ice age, or the slower receding of glaciers at the end of an ice age. The last maximum tilt was about 9,000 years ago, and we are moving toward a minimum tilt and lower seasonal temperature contrasts, which should occur in 12,000 years.

---

[20] Moore, P. 2015: *The Positive Impact of Human CO2 Emissions on the Survival of Life on Earth.*

Thirdly, Milankovitch showed that the Earth's rotational axis follows a slow, 26,000-thousand-year wobble, like a child's spinning top that is slowing down. This is because, like a child's toy top, our Earth is fatter than tall (but only very slightly). If the axis were an axle sticking out of the North Pole, it would draw a circle. Right now, our north star is where the axis is pointing, the star Polaris, and in about 13,000 years, at the other side of the circle of wobble, our north star will be the star Vega. Similar to axial tilt, this axial wobble affects the climate by increasing the contrasts between the seasons in one hemisphere, but what is different is that it will simultaneously decrease the seasonal temperature contrast in the other hemisphere.

Ice ages are predominately a northern hemisphere event because there is more land in the north and more water in the south. Land will cool down faster because it has a much lower ability to store heat; it has about 20% to 25% of the heat capacity of water. That means it takes only one quarter to one fifth of the energy to heat or cool land compared to a similar volume of water. When the Earth is farthest from the sun in its elliptical orbit, the whole planet receives less solar energy, and the planet is cooler. When the axis of the Earth is at a minimum tilt, winter is less intense (but still freezing at the poles), and the summer heat is weaker and less able to melt ice built up over the preceding winter. This sets up a positive feedback cycle for building thick ice sheets because in the summer, the white ice reflects more solar energy into space, further cooling the planet and allowing more ice to develop the following winter. On top of that, if the axis has wobbled over to a point in a specific direction, the northern hemisphere will have even cooler summers and milder winters — conditions favourable for glaciation.

Just as our planet orbits the sun, the solar system has its own 63-million-year orbit within our galaxy of the Milky Way. The solar system orbit is tilted at about 60 degrees from the plane of orbit of the galaxy, resulting in every 31 $^1/_2$ million years or so of the solar system orbiting into, across, and then out of the orbital plane of the galaxy. The galaxy is made up of four spiral arms of stars also orbiting at about 200 million years per revolution, which means that sometimes when the solar system passes through the orbital plane of the galaxy, it encounters one of these spiral arms of stars. This happens about once every 145 million years or so. Since some of these stars will be supernova (exploding stars) and the source of the cosmic rays that cause clouds on Earth, passing through some of them increases the cosmic rays we encounter on Earth. This amplifies the low cloud building effect in our atmosphere and cools the planet. Moreover, cosmic dust in one of these arms would further impede the sun's energy from reaching the Earth. Twice every 63 million years, the solar system's orbit will pass through the galactic orbit, and every 145 million years, one of those passages will be through a galactic spiral arm of stars. Each case is a cooling event caused by elevated or intensely elevated cosmic rays and cosmic dust. Either case could cause an ice age to be started, deepened or prolonged.

That's the end of the science of how climate change and ice ages could be triggered without the inclusion of carbon dioxide as a greenhouse gas. I am not saying this is proof that current climate change must be caused by natural variations in the amount of energy from the sun that reaches the surface of the Earth; at this point, just consider it a likely explanation for major climate change events in the past. The Milankovitch cycles and the rotation of the Milky Way are so slow-moving that they likely only bring about ice ages on Earth.

Sunspots are disregarded by the IPCC as only indicating small changes to the sun's energy output that are inconsequential to temperature increases seen on Earth in the last century. However, reputable scientists disagree and have calculated that half of the total temperature increase of 0.8°C seen in the last century could be directly from an increase in the sun's energy output heating the Earth with no other variables changed. In later chapters, we will delve into how much global temperature increase we have seen in the last century or so, and how confident we should be of those estimations. But for now, consider that as per Figure #5, between 1880 and 1980, there was a Grand Solar Maximum, where the number of sunspots doubled.

What is more significant about the increased energy output of the last century's Grand Solar Maximum is that the sunspots also indicate an increase in the sun's magnetic field, which extends well past the Earth, and which shields us from cosmic rays from the Milky Way. More sunspots mean a stronger shield and fewer cosmic rays, which result in fewer clouds and more of the sun's rays reaching the Earth. A lack of cloud cover on Earth due to a more active sun could easily explain more than the rest of the warming in the last century. Clouds are the Earth's thermostat and the IPCC forecast models, by their own admission, have difficulty dealing with clouds and ignore the connection of clouds to sunspots.

The 20th century was largely a Grand Solar Maximum period when direct heating output and decreased cloud cover could have substantially caused the reported global 0.8°C temperature increase. We have now entered an 11-year cycle of very low solar activity, and the 21st century is projected by some to be a Grand Solar Minimum, which should result in less direct heating and more cloud cover. It might get colder.

CHAPTER 3

History is the First Casualty

Chapters 1 and 2 examine how human-released carbon dioxide could add to the greenhouse gas effect up to a point beyond which there is little or no effect, and the solar science behind naturally occurring climate change. Yet, we are deluged daily by the global warming community with messages about human-released CO_2 causing "the hottest decade ever," "the hottest summer ever," "the hottest day ever in history," and that if we keep putting CO_2 into the air, it will just get worse. The media propagates this message when it knows-or should know-that there is a body of widely accepted scientific work that contradicts them: the Earth has generally been warmer than it is today, and that past major climate changes have been unrelated to atmospheric CO_2. Even the first report of the United Nations' mandated Intergovernmental Panel on Climate Change, released in 1990 and no longer posted on their website, acknowledged that the last 10,000 years were almost all warmer than today, and saw no correlation of those higher temperatures to higher levels of carbon dioxide. How do global warmers and the media ignore mountains of academic work that supports this history?

Simple. Just shorten history, manipulate historical records, or do both - whatever is convenient. In this chapter we will look at why global warmers keep shortening their historical database and, most importantly, what a longer view of human history tells us. In the next chapter we will look at the deceptive practice of selectively using historical temperature data.

Let's say you are a global warmer, and an internationally popular documentary has just publicized your group's claim that the Earth's temperature has been constant for a thousand years. But it recently shot up, and now we are living in the warmest temperatures of the millennia. Some busybodies, specifically Steven McIntyre and Ross McKitrick, inconveniently obliterate your conclusions by pointing out that during this same time interval, the Medieval Warm Period was considerably warmer than it is today, and the recently ended Little Ice Age was significantly cooler than it is today. There goes your theory that temperatures have been stable for a thousand years. Nevertheless, you can still claim the present are the warmest years in recorded history. Your response is to shrink the historical timeline to begin 200 years ago, which conveniently is just before the industrial revolution began to expand and enabled mankind to release CO_2 at a whole new level. Your new statement is that climate history dates back only the last 200 years because thermometers and records were unreliable before that.

Now, your skeptics point out that 200 years ago, the Earth was still rebounding from the historically well-established Little Ice Age, and that 100 years ago was the start of abnormally high solar activity. Therefore, it's obvious that the world was already on a natural warming trend. You respond by dismissing the sun as having any role in climate change and to now define only the last 100 years as recorded climate history. You justify this because the thermometers and records of the previous 100 years are much better compared to the last 200 years, plus there is an uncontested 0.8°C warming trend over a century that you can blame on CO_2 released by intensive global industrial expansion. But then there is a new problem with this shortened history: the people who grew up in North America in the 1930s remember this abnormally hot spell called the Dirty Thirties, which happened long before most of the human emissions of CO2 occurred. What to do?

Again, some of your global warmer colleagues shorten recorded history to 40 or 50 years with an asterisk and only accept temperature data gathered by NASA satellites or universities using approved thermometers. The global warmers' initial 1000 years of constant climate history that is used to justify a climate emergency has been shortened to 40 or 50 years. Still, the headlines of a runaway increase in the Earth's temperature and the hottest days in recorded history have not changed.

All the other disciplines of science combined have contributed so much knowledge, doubling every seven years now, that we don't have to restrict ourselves to only the temperature data collected and approved by the climate change community. We don't have to accept that the "hottest summer on record" means only the last 50 years. We can reconstruct climate history from much farther in the past, and we are about to do that. We are going to journey in time from the beginning of the Earth's modern climate era 12,000 years ago, named the Holocene. We'll see how climate change shaped modern man and not the other way around.

Most people have trouble with big numbers. To overcome that, we are going to imagine a male Homo sapiens born 12,000 years ago in southwestern Europe, and who will continue to live today and have the age of 100 years. Each one of his birthdays will mean the passing of 120 years on our Gregorian calendar; for every month elapsed on our man's clock, the Earth's clock will have elapsed 10 years. He will observe climatic events in the world and see how humanity adapted. Critically, he will show us that while there was a world temperature uptick in the 20th century, we actually live in a long-term global cooling trend. Let's call him Mr. Y. Dryas.

But first, some background. The long evolution from our closest ape relative, the chimpanzee, led to the present capabilities of our body and brain. Our body evolved to run long distances in pursuit of other animals for food, and our brain evolved to live in communal groups for mutual benefit and protection. The driving force for these Darwinian natural selections was climate change.

The earliest human ancestors split from the chimpanzees about four million years ago. This occurred during a period when the Earth's temperature was dropping from 5°C above today's temperature simultaneously with atmospheric CO_2 that dropped to its lowest level known, 180 parts per million, only 30 parts per million above the basic survival level plants need for photosynthesis. This meant that the forests of Africa disappeared, and savannas took their place. In the absence of trees, it was more advantageous to walk upright for both fight and flight. Then some 2.6 million years ago, the Earth's temperature dropped precipitously, during which the various Homo species became fully upright, and their tribal cooperation allowed them to run down even larger animals. The extra protein they got from running down big game enabled them to develop bigger brains, which takes a lot of protein. Bigger brains made better hunters with newly invented weapons, such as stone-tipped spears. Furthermore, group hunts made bigger protein prey possible. This virtuous circle of more protein equals faster, smarter, and more social Homo species continued until natural selection eventually resulted in Homo sapiens, about 300,000 or so years ago. Then 110,000 years ago, the Pleistocene Ice Age began, and about 75,000 years ago, Homo sapiens' frontal cortex brain development and highly socialized tribal life enabled speech. Quite possibly, the ability to speak triggered their migration from Africa to Europe and Asia, which occurred about the same time. [21]

Dryas' tribe migrated to Europe. Sea levels were 130 m lower than today, and Alaska and Siberia were joined by land. Beginning about 14,500 years ago, there was an abrupt warming of about

---

[21]   Walter, C: Last Ape Standing, (New York, Walker and Company, 2013), p 74.

12°C in a few hundred years, and the massive ice sheets began to melt. As the ice sheets melted, humanity's migration continued from Asia to North and South America. Sea levels rose 20m at a rate eight times faster than in the 20th century. Over the next 1,700 years, the temperature fluctuated wildly but trended downwards by 7°C or so, and 12,800 years ago, they plunged abruptly back to the levels of the ice age. This sudden and global return of glaciation conditions is called the Younger Dryas [22], named after the flower that bloomed in Europe during these cold times. It was also when our man Dryas was born.

There is a bit more technical background to review also. Most people, encouraged by media reports and local experience, equate hot weather with dry weather, and cold weather with damp weather. On a global scale, the opposite is true. More heat causes more evaporation. There is more ice-free water surface area for evaporation, and warmer air can hold much more water as humidity. Globally, warmer means wetter, and warmer and wetter generally means a greener, lusher planet. Colder conditions mean more water is locked up as ice, some of which may cover open water. There is less heat for evaporation from open water, and the air has a much lower capacity to hold water. Globally, colder means drier, and colder and drier generally means browner and dustier.

In the narrative below, unless otherwise indicated, the temperatures quoted are taken from Greenland ice core samples, and the atmospheric carbon dioxide levels are taken from Antarctic ice core samples. Ice cores captured the oxygen in the air when the ice or snow was formed. Oxygen can come in a heavy variety, Oxygen Isotope 18 (with 8 protons and 10 neutrons); or a light variety, Oxygen Isotope 16 (with 8 protons and 8 neutrons). Water vapour that is made up of two hydrogen and one heavy oxygen (isotope 18) will condense and fall as rain quicker in cold temperatures than water vapour made up of two hydrogen and

---

22   Plimner, I: *Heaven and Earth*, p 41.

one light oxygen (isotope 16). In cold times, the heavy oxygen water is less prevalent in polar snow because it was rained out of the atmosphere before it got to Greenland. A higher ratio of Oxygen Isotope 16 (lighter) to Oxygen Isotope 18 (heavier) in the Greenland ice cores means it was colder when that snow fell. It is an indirect measurement (a proxy) of temperature, but when tied to a direct temperature, such as today's Greenland temperature, it is very useful.

Whether you are a global warmer or a heretic, there is an easily recognizable limitation to using ice core derived temperatures as a global temperature. It is assuming that the entire globe had temperature changes of similar magnitude and direction to that one spot where the snow fell. Ice core derived temperatures gain credibility as global indicators when other branches of science can identify simultaneous similar temperature changes in other parts of the world by various means. This interdisciplinary science approach can help complete the puzzle. This is a far more comprehensive approach to reconstruction of historical temperatures than the method used in the initial global warming claim that the Earth's temperature has been constant for a thousand years (we will also discuss that in a later chapter). I chose the Greenland ice core data for temperature as they are widely available in the public domain. Carbon dioxide can be directly measured from air bubbles trapped in the ice, and I chose the Antarctic ice core data for CO2 levels as they are widely available in the public domain.

An Uncertain Beginning: Age 0 to 4 (Earth clock from 12,000 to 11,500 years ago)

It's very cold, and Dryas' parents are worried about having enough food for another mouth, how to keep him warm, and how far south the tribe will want to go to avoid the advancing ice sheets. Their migration is interrupted by the most dramatic climate change event known to history when the temperature increases over 10°C in about half of a century (Earth clock). This is over 20 times the rate of temperature increase that will alarm global warmers in the 21st century, and we don't have a widely accepted explanation for it. [23] Allowing for a few centuries of lag time for ocean degassing caused by the warming water, the carbon dioxide levels in the atmosphere during this warming have increased from 240 to 260 parts per million. There are about 2 $^1/_2$ million people living around the globe, with technology still in the Stone Age.

A Happy Childhood, Adolescence, and Young Adulthood: Age 5 to 29 (Earth clock from 11,500 to 8,500 years ago)

This is an excellent time to be a human. The Pleistocene epoch is over, and the warming trend, called the Holocene Warming, continues. Forests expand into the grasslands, and grasslands expand into the deserts. Temperatures are about 2°C warmer than the 20th century, and atmospheric CO2 is stable at 260 parts per million. The Stone Age occupation of nomadic hunter/gatherers is mostly left behind as the agricultural revolution starts in Iraq and introduces a new cultural innovation—the village. There are now about nine million people in the world.

Early Mid-life Crisis: Age 30 to 53 (Earth clock from 8,500 to 5600 years ago)

Dryas' good life gets interrupted with a 500-year speed bump in human prosperity. Unexpectedly, the temperature dropped about 2 ½°C and then popped backed up again to the same level. That's a 1°C temperature swing per century, in each direction. Allowing for ocean absorption and degassing lag time, the atmospheric CO2 dropped five parts per million and then bounced back five parts per million. This temperature drop and recovery may have been caused by the rupture of natural dams in Canada which

---

23   Easterbrook, D. 2012: *The Intriguing Problem of the Younger Dryas – What Does it Mean and What Caused it?*

were holding back massive lakes of meltwater from the receding ice sheets. This released cold freshwater that floated on top of the salty oceans for 500 years until they were mixed with the warmer salty water and temperatures then rebounded. In the European Alps, the treeline descended, and vegetation changed in response to the cooler climate. This cold spell caused drier conditions in North Africa (Egypt) and East Africa, resulting in droughts. Again, mankind adapted to climate change. In Mesopotamia, the drier conditions may have been the necessity that gave rise to the invention of crop irrigation techniques. Others in Turkey descended from the highlands to the warmer lowlands in a vast fertile plain, which was 100 metres below sea level. Around 7600 years ago, plate tectonics ruptured a protective ridge that opened the Bosporus Straight, and great flooding of the inhabited low plain ensued, creating the Black Sea and the biblical story of the great flood. Once the temperatures rebounded, the Holocene Warming period continued to 5600 years ago. The temperature in the Arctic was up to 3°C higher than today, and the Sahara was wet enough to support herds of animals. [24] Passes in the Swiss Alps that were heavily glaciated in the 20th century are open to human foot traffic. The world's population is now 34 million people.

Maturity and Potential: Age 54 to 79 (Earth clock from 5600 to 2500 years ago)

Temperatures are fluctuating within a window of today's level to 2 ½ °C higher than today, and the final climb of CO2 in the atmosphere happens from 260 to 280 parts per million, the value we use as the pre-industrial level. Dryas observes that this is again about the same temperature swing as in his Early Mid-life Crisis phase. Due to the much larger human population and organized governments, the science of archeology now helps in delineating

---

[24]   Plimner, I: *Heaven and Earth*, p 49.

the climate change effects. For the first time in history information will be recorded by the invention of writing.

From 5600 to 3500 years ago, there is a cooling period down to today's temperature. Droughts hit the cradles of civilization hard. The Sahara is deserted as it transitions into a desert. China, North America and South America also experience desertification. The Middle East Empire of Akkadia is partially abandoned and falls, Greece becomes deforested, and written records detail poor crops. Alpine Europe reglaciates. West Virginia caves preserve evidence of long droughts. Kilimanjaro ice shows an African drought. Humanity, in general, transitions into the Bronze Age.

The next 300 years (3500 to about 3200 years ago) are the warmest of Dryas' life. There is a temperature spike that is at least 2 ½ °C warmer than the 21st century, well past the range that the 21st century global warmers claim is catastrophically critical and must be avoided at any cost. CO2 levels have reached 270 parts per million. This is the Holocene Maximum, and across North Africa, the Middle East, and China empires, kingdoms and dynasties flourish. In Dryas's extended community, some return to Scandinavia to reclaim abandoned farmland.

The final 700 years (3200 to 2500 years ago) is again a cooling period, but milder and still at least 1°C warmer than today. This is now the Iron Age, also entering the Greco-Roman Age, and the world's population has grown to 150 million.

Golden Years: Age 80 to 87 (Earth clock from 2500 to 1485 years ago)

This is an important time in Dryas's life because civilization is advanced, and the foundations of life in the 21st century are being laid. The population of the world grows to 210 million. Contemporary writers record much of the natural environment, and much can be inferred about climatic conditions. For example, it is the peak of the Roman Empire, and in Scotland they are wearing togas and growing grapes, which indicates a warmer climate. Scholars will debate whether the world was 2°C warmer (Greenland ice cores) or up to 6°C warmer using other indirect measurements, but to enter this discussion would be to open the forbidden exercise of hair-splitting data collection, data accuracy, and global applicability. Rather than debate the accuracy of the temperatures derived by inaccurate proxy methods, it is better to view the climate by what happens from the temperature increases. What Dryas sees is that the Romans are growing olives in the Rhine, which is warmer than in the 21st century. The Vikings are using an ice-free high mountain pass at Lendbreen in southern Norway to get to high elevation summer pastures. In the 20th century, this pass was ice-covered.

Dryas makes another important observation that will later be ignored by the 21st-century global warming community: there is a thick forest growing in northern Quebec, Canada, 130 km north of the 20th-century tree line. This is proof preserved in a peat bog that North America also shared in the global Roman Warming Period and that the late 1990s could not have been the warmest global temperature in a millennium.

Call 911! Ages 88 to 99 (Earth clock from 1485 to 170 years ago)

After the wonderful Roman Warming, there were about 365 years of hell, from 1485 to 1120 years ago. The temperature dropped to about 1°C cooler than in the 20th century. The sudden cooling caused crop failures and famines in Europe. 25 million people died of the bubonic plague. Ice was reported on the Nile and the Black sea. Glaciers grew in North America, and droughts weakened and eventually destroyed the Mayans in South America. It was the Dark Ages, or more accurately, Cold and Dark Ages, and it was global.

The Medieval Warm Period followed for the next 400 years, from 1120 to 720 years ago. It was not as warm as the Roman period, but still up to 2°C warmer than today, depending on location. Dryas saw agriculture expand northwards in Europe, their food was plentiful, and they were wealthy enough to start the Crusades. He saw the Vikings establish farms in Greenland and a settlement in Canada, which they called Vinland because they discovered grapes growing there. The English grew grapes again. Economic prosperity returned to Europe where cathedrals were built, and Chinese and Middle East empires flourished also. Intercontinental trade started. Vegetation, pollen, and stalagmite studies, as well as written records, analyzed in over 100 studies, confirm it was a global event equal to or greater than the global warmers' climate emergency of the 20th and 21st centuries.

We previously discussed that a high level of sunspots results in decreased cosmic rays hitting the Earth's atmosphere. Carbon-14 is produced in the atmosphere by contact with cosmic rays, so when less Carbon-14 is found in tree rings, this indicates the sun was more active during that tree ring growth year and is reducing the amount of Carbon-14 in the atmosphere. From tree rings, we know the Medieval Maximum of solar activity accompanied the Medieval Warm Period, and there were no large-scale human carbon dioxide emissions. Once again, Dryas had a good life, until, again, he didn't.

The Little Ice Age began 720 years ago, and it was linked to lower solar activity. Some of this was detected by the Carbon-14 indirect (proxy) method, and some from direct astronomical observations with the invention of the telescope. Between 740 years ago and 170 years ago, there were four recognized periods of lower solar activity: The Wolf, Sporer, Maunder, and Dalton Solar Minimums, and there were no solar maximums. This is also a highly documented period of human history, as at the beginning of it, there were 390 million people on Earth, and at the end, approximately 1.25 billion. The temperature decline from the peak of the Medieval Warm Period to the coldest of the Little Ice age was just a few degrees, and only slightly cooler than the 20th century. Therefore, it wasn't quite an ice age, and mankind did reasonably well due to rapidly advancing technology. About

400 years ago, the forests in northern Quebec, Canada that were started in the Roman Warming Period are wiped out in a forest fire, and it is too cold for them to regenerate.

This brings Mr. Dryas contemporaneous with Svante Arrhenius and his famous prediction that anthropogenic carbon dioxide emissions will cause greenhouse gas effect temperature increases on Earth. Of course, Arrhenius did not know that just 50 years later, a new, massive increase in solar activity would begin, the Modern Solar Maximum.

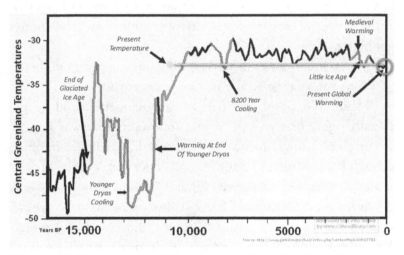

Figure #6: Temperature Graph from Greenland Ice Cores

Century Celebration: Age 100 (Earth clock is today)

Reporter: Mr. Dryas! Happy 100th birthday! May I interview you about your most remarkable life? 12,000 years from cave to condo!

Dryas: Of course! I have lived through a lot, and to share that is the purpose of me being here.

Reporter: When was the best time to be alive? When was the worst?

Dryas: I'll start with the worst. As you know, I had a tough beginning 12,000 years ago, but when it warmed up, life was good. When it was warm, we generally had good crops and enough food, and when it was cooler, we didn't. The best time to be alive is today, not because it is warm–it is actually cooler–but because we know so much. We also know what we don't know, but we have a way of figuring it out through the scientific method, which is important. The worst time was the Dark Ages because it was our ignorance that was hurting us the most. It's not that we were not smart back then, but the political leaders, like many before them, did not want us to think for ourselves. Instead, they just wanted us to follow their directions without question. Getting rid of that political control over intellectual life was critical, and then things got better.

Reporter: So, democracy was the big turning point for a better life? One vote, one man, and no one is above the law?

Dryas: No, young lady. Although democracy and all the human rights that go with it are important, that was not the turning point for humanity in terms of a better life, meaning enough food, adequate housing, good education, excellent health care, and a clean environment. When I was born, there were only 2 $^1/_2$ million of us, but we all had to think for ourselves to survive, and we shared that thinking to choose the best path forward. Yes, mistakes were made, and without writing, we forgot some important lessons, but we were all welcomed to voice our thoughts. Now that there are 7 $^1/_2$ billion of us, and life is so specialized, we have lost that ability to contribute, and seemingly, have also lost the *right* to contribute. In the worst cases, small groups of people have taken over the discussion of critical issues that affect us all, and they do not want the rest of us to think for ourselves.

Reporter: Mr. Dryas? I think you had a senior's moment just now. You rambled on, and I didn't understand you.

Dryas: Miss, I think you had a millennial's moment. You zoned out and missed my meaning. I'll spell it out. You see, when you allow a small group to control any critical issues that affect your life, the outcome can only be as good as the talent in that small group, possibly further limited by the prejudices of the leadership of that small group. What if they are wrong? If you allow input and discussion from a much larger, open, and engaged intellectual community, the interactions within that community ensure you will have a result that is better than any individual in that larger community, and much better than the smaller controlling group. The turning point in human well-being in the last 12000 years was when past generations stood up and challenged, "Dare to know! Have the courage to use your own understanding." Those larger communities took over intellectual life from the small control groups and gave us the Age of Enlightenment, the freedom of independent thought, which led to all the other freedoms we enjoy in modern democracies.

Reporter: That does sound important, but in many places, the world is still controlled by small groups looking after themselves. Can you give me a contemporary example that concerns you?

Dryas: Look at all the warm and cooler trends during my lifetime, and then during that same time, look at the big swings in solar activity and the relatively stable carbon dioxide content of the air. Where do you see a correlation? On which basis would you forecast future climate? And who would you entrust with conducting that forecast? What if, based on that forecast, the government decided that drastic changes to you and your children's lives were required, and when you asked questions, the answer from a small controlling group was, "It's settled. If you don't do as we say, Armageddon, in the form of a drastically warming Earth is arriving shortly." For 90% of my lifetime, the world has been warmer than today, and I rather enjoyed it.

Reporter: I'm sorry, Mr. Dryas, I admit I was initially dismissive, but now that you have come out as a Climate Change Denier, I can see that you are part of the problem and not part of the solution. But the rules in my media game are that I must condense your 12,000 years of wisdom into five seconds or 280 characters. Can you do that?

Dryas: If we want to solve our environmental problems, we cannot accept control from a small, unaccountable political body, or we could lose much and gain nothing. The solutions will require us all to be knowledgeable of the science and courageous with the truth.

Reporter: OK, boomer.

Dryas: And I'm a heretic, not a denier.

CHAPTER 4

There is No Climate Emergency; There is a Crisis in the Intergovernmental Panel on Climate Change

Here is a quick review of this book thus far:

- Chapter 1: By far, the most important greenhouse gas is water vapour, which contributes to 90% of the greenhouse gas effect and cannot be controlled by humans. Carbon dioxide may have little or minimal greenhouse gas effect above the current levels as the amount of C02 needs to be continually doubled to get the same temperature increase. This and convection currents of water vapour are why the Earth has never had a runaway greenhouse effect since multi-cellular life began, even with 12 times the CO2 in the atmosphere. CO2 is healthy for plants and not hazardous to humans; it is essential for all life and cannot acidify the oceans.
- Chapter 2: Solar activity variations can cause climate change on Earth by changing the amount of short-wavelength

radiation sent to the Earth and changing the cloud cover that reflects the radiation into space. Sunspots track solar activity variations, and more sunspots mean global warming conditions. NASA predicts that the next 11-year solar cycle will be the weakest one in 200 years. Some geophysicists are predicting it will mark the beginning of a new possibly decades-long Grand Solar Minimum. This is not factored into the IPCC climate forecast models.

- Chapter 3: When the Younger Dryas cold period ended about 11,500 years ago, the global warming trend caused CO_2 concentrations in the atmosphere to increase, not the other way around. In the last 10,000 years, the Earth has been warmer than it is today, 90% of the time. During this time, there have been several natural warming and cooling cycles greater than the 20th century, during which the CO_2 level in the atmosphere was relatively constant.

The statements in the chapters above are not my own, nor are they personal preferences or interpretations. They are scientific observations taken from the classic perspective of "Standing on the Shoulders of Giants." [25] They are based on centuries of incremental scientific progress made by many scientific giants since the Age of Enlightenment. To successfully repudiate these observations, you might be a giant yourself, which needs to happen and does happen.

Aristotle observed that a rock will fall faster than a feather, and thought it was because the feather was lighter. Galileo Galilei experimented and found that Aristotle made an error, that any

---

[25] Many attribute this phrase to Isaac Newton, who said that if he could see further, it was because he was standing on the shoulders of giants. Newton was quoting Bernard of Chartres from the beginning of the 12th century, who was telling his students they could understand more not because they were smarter, but because they had the benefit of smarter people before them.

two objects of different densities but the same size and shape would fall at the same constant rate. Isaac Newton proved that the constant rate Galileo measured was unique to Earth, as gravity is an attractive force between two bodies based on the relative mass of each of the bodies. Therefore, he expected gravity would be weaker on the moon. Then Albert Einstein worked out that mass is just another form of energy—it's all the same stuff—and time is just another dimension of space. Stephen Hawking followed up with a theory that this mass/energy/time/space stuff was all started with a Big Bang, and then the Barenaked Ladies put it to music.

Science is never wholly settled because knowledge is never complete. People who say the science is settled do so because they want to end the debate. That's not how science works. Arguably, Isaac Newton was the smartest human being that we know of, and yet, his work was significantly debated by other intelligent people building on his formidable body of work (he also invented calculus and the post office). It is these giants, and many others, who form the current understanding of our universe, and I am more secure standing on their shoulders than relying on the Intergovernmental Panel on Climate Change (IPCC). Why? Simply, the IPCC is not a scientific organization conducting climate research; it is part of a political organization seeking political consensus. A new friend of mine, who was involved in politics for over a decade, advised me that in politics, facts do not matter. What matters in politics is what ideas get accepted as facts, and it is of secondary importance whether they are true. I naively hope this is a viewpoint jaded by his bitter experience, as it is the opposite of science, which seeks the truth regardless of the politics.

What we are going to look at in the first part of this chapter is that the IPCC does not produce scientific facts; it produces opinions of scientists influenced by politics. Secondly, I'll provide a heretic's interpretation of the same recent historical data used by the IPCC and conclude that there is no climate emergency. In the last part of this chapter, we are going to look at how the political colonization of climate change science has resulted in a credibility crisis at the IPCC.

The Scientific Method vs. the IPCC Method

Galileo Galilei is considered by many to be the father of modern science; he gave us the telescope, the thermometer, and interesting educational use of the Leaning Tower of Pisa. I think his greatest contribution was laying the foundation to something you learned early in your academic life—the Scientific Method.

During the Dark Ages, the average person in Europe knew nothing of the Roman and Greek scientific knowledge that was previously taught at Greco-Roman schools. The church combined religion, politics, and science into a tightly controlled one-window service. Because Aristotle's scientific view that the universe revolved around the Earth fit nicely with the church's religious belief that God created man in His image and therefore man was the centre of the universe, the only science the Church wanted you to know was what originated from Aristotle. He theorized the Earth is the centre of the universe; lighter objects fall more slowly than heavier objects; and water, earth, wind and fire are the four elements that make up the universe. Galileo challenged these concepts with experimentation, and a contemporary of his, Francis Bacon, built on that work by codifying the scientific method. Bacon used Galileo's process of observation and reasoning to advance scientific inquiry. Interestingly, Francis Bacon was a lawyer, but then again, so is John Cleese. You may have been taught that there are five, six, or even seven steps to the scientific method, but that is normally just spreading out the basic five steps, which are:

1. Make an observation.
2. Ask a question (This is where an extra step is often placed: research the question.)
3. Formulate a hypothesis.

4. Conduct an experiment to test the hypothesis.
5. Analyze the data, and draw a conclusion (This is sometimes split into two steps).

I will engage some literary licence and speculatively fill out the five steps of the scientific method as I think the IPCC might do:

1. Observation: The climate is changing; the Earth is getting warmer.
2. Question: Is human activity contributing to this global warming?
3. Hypothesis: Human emissions of the greenhouse gas carbon dioxide are contributing to global warming.
4. Experiment: Solicit and evaluate multiple computer simulations from experts on climate change.
5. Analyze the data and draw conclusions: The hypothesis is confirmed; multiple computer simulations confirm that human emissions of carbon dioxide are substantially contributing to global warming.

If following the steps of the scientific method qualifies as science, and the IPCC could easily fill out the above themselves, why do I assert their methods are not scientific? It is because of step 4. The computer simulations conducted on behalf of the IPCC do not qualify as experimentation, and therefore, the process fails to pass the basic test of scientific inquiry. This is not my original thought; I have borrowed it from one of the most influential early supporters of the IPCC, who later became one of its most prominent critics, Margaret Thatcher. The Iron Lady was not for turning, except when it came to the IPCC.

Francis Bacon did not have to worry about whether computer simulations could be used as a scientific experiment, but we do. Computers can speed calculations that would otherwise take a prohibitively long time, and they can simulate conditions that might otherwise be unsafe or impossible to observe. There are two very logical tests they must pass to qualify as an experiment. Firstly, computer simulation must be validated, and secondly, it must be verified. Validation is confirmation that the right mathematical equations were chosen to simulate the conditions of a physical experiment. Verification is just confirming that the equations in the computer simulation were solved correctly.

Engineers use computer simulations all the time, and I did extensively in my career. Often, the test of validation of future predictions was met by setting the simulation clock back several decades, inputting the assumptions, and then running the simulation to see if it replicated known history. The IPCC scientific contributors do the same, and we call this a history match. Even if you got a good history match, you had to take the prediction of future events with a grain of salt. Often, several different sets of input data would result in the same history match, and the computer technicians were very skilled at obtaining the history match by whatever assumptions were necessary. If questionable assumptions led to a good history match, the computer technicians had an expression for this: garbage in equals garbage out, and the results would be discarded, validation still failed.

Let's look at a simple example of how a simple mathematical model, which forms the basis of all computer simulations, can easily result in a wrong answer. In January 1991, the *Dick Davis Digest* asked a serious question: How many Elvis Presley impersonators will we have by 2010? So, they made a serious mathematical model to predict it. They input the data:

- In 1960, there were 260 Elvis impersonators
- By 1970, there were 2,400
- In 1990, there were 14,000

Their mathematical model predicted that by the year 2010, one in four people in the world would be an Elvis impersonator, a total of 1.7 billion Elvises. This teaches us three important things about mathematical modelling and computer simulations. One: even if you have the correct equations in the model, without built-in constraints to limit the equations, the model can give false results. For example, a limitation is that the world economy can only support so many Elvises. Two: just because part of the model works, doesn't mean the entire model does. For example, Dick Davis made a very accurate prediction of the world population 19 years in advance. Therefore, one could claim the model was 50% accurate in predictions, excepting, of course, the prediction that mattered was the failure. Three: just like *garbage in* equals *garbage out, humour in* equals *humour out.* Most importantly, the Elvis example is to illustrate that like humour, *politics in* will equal *politics out.*

Now, let's look at the IPCC global climate models. The IPCC admits they cannot simulate clouds very well, and a 1% change in cloud condition could explain all the global warming in the 20th century. So, already the climate models fail to be validated as a scientific experiment. The IPCC computer simulations either minimize or ignore the extra energy that an active sun can provide, which directly heats the Earth to a 0.45°C difference from a lazy sun to an active sun. Hence, it fails validation a second time because it willfully ignores a valid consideration. It is very difficult for anyone to mathematically simulate global water evaporation combined with turbulent convection, which accounts for 70% of the heat loss from the surface of the Earth. This is the third failure.

The models also predict that warming will be more severe where sunshine is the strongest and will be detected as hot spots in the lower atmosphere of the tropics. In reality, the Arctic is warming faster than the tropics, and the Antarctic has mostly stable temperatures. I realize the scientists contributing to these models are highly qualified PhDs, but the supreme law of science, the second law of thermodynamics (entropy), states that energy will move from hot to cold. Entropy predicts that additional heat energy in the tropics will move to the poles, mostly by water, due to its very high heat capacity. Warm currents cannot reach Antarctica because of the circumpolar current there, but warm currents can reach the Arctic. The Arctic warming faster than the tropics should not have been a surprise; it should have been the first expectation, so there is something else wrong with the models.

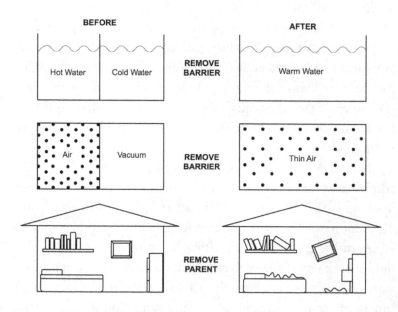

Figure #7: The Second Law of Thermodynamics.

The level of disorder (Entropy) increases naturally, meaning that the heat and concentrations will balance out unless constrained.

Beyond the problem of what calculations are required in a global climate model to make it valid as science is the even bigger problem of where the historical temperature data comes from to run the models. The first big issue is how do we determine the true average temperature of the Earth, now and in recent history? It is very problematic. Recall we have discussed that water can hold five times the heat as a similar volume of land, and ocean currents deliver that heat to various parts of the world (like the Arctic) and not to others (like the Antarctic). This means that 70% of the sun's energy, which falls on the oceans, is distributed unevenly around the world. Taking the temperature of all the oceans in the world and blending it with the temperature of all the land in the world to estimate one global average temperature is a big problem, and the method used to accomplish that will affect the results of the climate change simulation models.

A second historical temperature factor is the urban heat island effect, which can raise the temperature of what was previously a farmer's field and is now an urban area. Cities generate heat from buildings, industry, and automobiles and can store this heat in pavement and brick.

A third historical temperature factor is that at the end of the 19th century, the thermometers in everyday use had much less accuracy than today, resulting in the potential error of older thermometer readings being equal to the claimed global temperature increase since then.

As promised, we will not venture too far into this hair-splitting debate about historical temperature data. Instead, we will just point out that one way to make the IPCC computer simulations match history is to carefully select which temperatures you use as history (which thermometers to believe), how you may or may not correct them (for example, the urban heat island effect), and what statistical manipulations you exercise to come up with a blended global average temperature. If your objective is to make a forecast seem credible by achieving a good history match, then carefully selecting historical temperature data assists your cause. Too much of this behaviour can result in garbage in.

The IPCC global climate change models fail multiple times to be validated as scientific experiments. Therefore, they do not meet the criteria of the scientific method, and they do not constitute scientific fact. They are in aggregate the best estimate of the future climate of Earth made by scientists working on that subject under the umbrella of the IPCC. The models reflect the biases, opinions, and prejudices of the IPCC.

With all these shortcomings of the IPCC climate simulation models, one wonders about the accuracy of the forecasts. The IPCC does not rely on just one model, but they aggregate the results of many models run by separate groups as a basis of their forecasts. An independent analysis [26] of these models was published in 2014, comparing the forecasts using the same average global surface temperatures database as the IPCC (known as HadCRUT4). It found that the aggregate of 108 IPCC models used:

- For the previous 30-year period (1985 to 2014), the predicted global temperature increase was 0.75°C. The actual measured global temperature gain was 0.5°C.

- When the simulation window was narrowed to the previous 20 years (1995 to 2014), the predicted average global temperature increase was 0.54°C. The actual measured global temperature gain was 0.2°C.

- With a simulation window of just 10 years (2005 to 2014), the IPCC computer simulation forecast average global temperature increase was approximately 0.2°C. The actual measured temperature gain was nil.

Up to the year 2014, over the previous 30-year period, the IPCC models overestimated the global temperature increase by a factor of 1.5. Over the last 20-year period, they overestimated the global temperature increase by a factor of about 2.5. Finally, over the previous 10-year period, the forecasts missed that there was no global warming. These are the forecasts our governments

---

26 Knappenberger, P.C., Michaels, P.J. 18Dec 2014. CATO.org; AGU: *Quantifying the Mismatch Between Climate Projections and Observations.*

are asking us to accept as fact, as the foundation of international climate change agreements for the next 30 years, and as the justification for the permanent restructuring of our economies.

By the way, in 2010, there were an estimated 85,000 Elvis Presley impersonators worldwide. That forecast was off by a factor of 20,000, but it was funny.

A Heretic's Interpretation – There Is No Climate Emergency

IPCC climate forecast models may be run by scientists, but the results are not scientific fact, so let's not use them. Instead, let's use the data collected by the IPCC and published in the public domain to draw our own conclusions by using simple logic.

The public domain files of the IPCC state: "In fact, the absolute concentrations are not especially important, as the temperature response to increasing CO_2 concentration is logarithmic—a doubling from 500 to 1000 ppmv would have approximately the same climatic effect." [27] This is consistent with the imaginary trial of CO_2 we had in Chapter 2. The IPCC is confirming they agree that whatever level of greenhouse gas effect we have seen in the last century, it will take twice as many CO_2 additions to the atmosphere to get the same temperature increase.

For the sake of illustrating the impact of the logarithmic relationship, let's be diplomatic and take a 50/50 approach. Let's assume that of the 0.8°C temperature increase seen since the beginning of the last century, 50% was caused by human emissions of CO_2, and 50% was natural causes (mostly solar variations). This would result in a 0.4°C temperature increase caused by a 120 parts per million gain in CO_2. You would then need to add 240 more parts per million CO_2 to the atmosphere, for a total of 640 parts per million, to get another 0.4°C temperature gain. Now, let's look at the IPCC's most recently posted (as of this writing)

---

[27]   www.IPCC-data.org/guidance on the use of data/Consistency and Reporting.

forecast for atmospheric CO_2 concentrations. [28] The forecast only goes to the year 2100:

- Three of the six forecasts do not make it to 640 parts per million before 2100.
- The other three reach 640 parts per million between approximately 2060 and 2080.
- The average of all six forecasts results in 640 parts per million CO_2 in about 2080.

From the IPCC public domain files, we are likely 60 years (two generations) from hitting 640 parts per million CO_2 in the atmosphere, where the temperature is 0.4°C hotter than today and 1.2°C hotter than the pre-industrial level. If you are worried about a post-2080 temperature increase of a further 0.4°C caused by CO_2 in the atmosphere, we would need a further 480 parts per million (doubled again) CO_2 for a total of 1020 parts per million to reach a total temperature increase of 1.6°C since pre-industrial times.

If diplomacy fails and the 50/50 split is disagreeable on both sides, then you have a range of outcomes for the year 2080, with 640 parts per million CO_2 in the atmosphere. If you assign 100% of the last century's temperature increase to the CO_2 increase, then the temperature will be 0.8°C hotter than today and 1.6°C hotter than pre-industrial times. Furthermore, if you attribute 100% of the last century's temperature increase to natural causes (solar activity), 2080 will have the same temperature as now.

---

[28] www.IPCC-data.org/data: simulations/CO2: Projected emissions and concentrations/Figure 1.

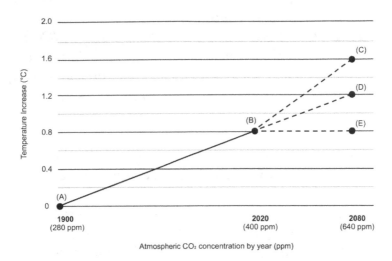

Atmospheric CO₂ concentration by year (ppm)

Figure #8: Where is the Climate Emergency?

Range of Possible Temperature Outcomes based on IPPC CO2 Predictions, CO2 Sensitivity and Warming to Date.

(A) Baseline temperature from early industrial days, CO2 is 280 ppm.

(B) Global temperature is 0.8°C warmer, and CO2 has increased to 400 ppm, an increase of 120 ppm.

(C) IPCC average prediction is about 640 ppm CO2 in 2080, which is a doubling of the 120 ppm gain from 1900 to 2020. If all the temperature increase from 1900 to 2020 was due to CO2, then in 2080, the temperature should be 0.8°C higher than today or 1.6°C higher than 1900. CO2 sensitivity is approximately 1.4°C for each doubling of CO2.

(D) If 50% of the temperature increase from 1900 to 2020 was due to CO2, and the rest was natural warming, then in 2080, the temperature should be 0.4°C higher than today or

1.2°C higher than 1900. CO2 sensitivity is approximately 0.7°C for each doubling of CO2.

(E) If all the temperature increase from 1900 to 2020 was natural (CO2 sensitivity is 0°C for each doubling of CO2), then in 2080, the temperature should be the same as today or 0.8°C higher than in 1900 (unless there is more natural warming or cooling).

There is a second very relevant release by the IPCC that further undermines the declarations of a climate emergency [29]:

The rate of warming over the past 15 years (1998–2012; 0.05 [–0.05 to +0.15] °C per decade) is lower than the trend since 1951 (1951–2012; 0.12[0.08 to 0.14] °C per decade).

That last quote was taken from the most recent IPCC Assessment Report (AR5) of 2014. However, Working Group 1 (The Physical Sciences Basis) submitted their section a year earlier, in 2013, and the data they had was up to the end of 2012. If you found it to be a bit confusing, that's because it was written to hide an inconvenient truth. The quote above is confirmation from the IPCC that from 1998 to 2012, global warming ceased, and it occurred during a time described in the report as having the highest anthropogenic greenhouse gas emissions in history.

Some heretics call this period the Global Warming Hiatus, and the data support a cessation of global warming. The statement from the IPCC is contrived to be misleading because it compares two trends, one from 1951 to 2012 with a warming trend of 0.12°C per decade, and another from 1998 to 2012 with a warming trend of 0.05°C per decade. Note, each trend ended in 2012. A better

---

[29] IPCC. Assessment Report 5, WG1, Climate Change 2013 The Physical Sciences Basis, p 37.

comparison would be from 1951 to 1998 with a warming trend of 0.13°C per decade, and from 1998 to 2012 with no warming trend at all, possibly a slight global cooling.

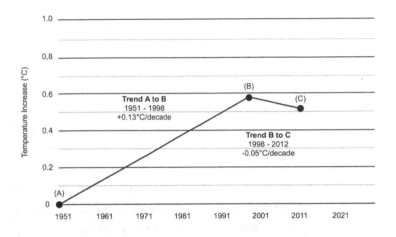

Figure #9: Is This the Climate Emergency?

The preferred IPCC database shows no global warming from 1998 up to 2012, the data cut-off for IPCC AR5.

Using the same time intervals and sourcing the public domain database, as the IPCC used above, the 1998 temperature (Point B) was about 0.6°C warmer than 1951 (Point A). The global average temperature in 2012 (Point C) was 0.1°C cooler than 1998. That degree of accuracy in global average temperature is probably not possible. However, the IPCC statement "a much smaller increasing linear trend" obscures that between 1998 and 2012, there was no global warming.

Later in this chapter, we will examine why the IPCC's preferred global temperature database, called HadCRUT4, should be viewed with caution. Even with that caution in mind, the worst-case scenario presented by the IPCC files is a 0.5°C (3%) global temperature increase over the last 61 years, with the final 15 years flatlining. That's not a climate emergency; that's the rate of return on my retirement investment portfolio.

From the heretic's interpretation, we have cause to ask if we might have already hit and surpassed the peak rate of global warming (for clarity, global warming may still be occurring but at a much slower rate):

- There may already be enough CO_2 in the atmosphere, and there is very little heat radiation in the correct wavelength (14 to 16.5 micrometres) for any further increase in the greenhouse gas effect. This can also be interpreted from NASA satellite data.
- Because of the logarithmic relationship between CO_2 and temperature gain, we need to keep doubling the amount of CO_2 to get the same temperature increase.
- The last 540 million years of the Earth's history says there has never been a runaway greenhouse gas effect caused by CO_2. This may be because convective cooling has a more prominent role to play than the computer simulations may predict.
- From 1700 to 1950 there have been 23 solar cycles with an average peak of 165 sunspots, and from 1951 to 1998 there were five solar cycles with an average peak of 220 sunspots. From 1999 to 2019 there have been two solar cycles with an average peak of 145 sunspots. NASA predicts we are now entering the weakest solar cycle in 200 years, and this might be the beginning of a decades-long Grand Solar Minimum. During the Little Ice Age there were four consecutive Grand Solar Minimums.

Where is the climate emergency? Consider the two statements below. The first is paraphrasing from the Summary for Policy Makers in the IPCC Climate Change Report 2014 Synthesis

Report (which is the latest available). The second is a connect-the-dots heretic's logic from the various technical backup documents posted by the IPCC in support of the Synthesis Report. Note that the statements are not mutually exclusive.

> IPCC: Computer simulations show that if we have drastic cuts to human-emitted atmospheric CO_2, the temperature rise of the Earth will probably be limited to between 1.5°C and 2°C from pre-industrial times by the year 2100.

> Heretic: The logarithmic relationship between temperature and CO_2 concentrations combined with the temperature gain observed in the 20th century means that if humans continue to emit CO_2 into the atmosphere in a business-as-usual scenario, the temperature of the Earth will probably be limited to between 1.5°C and 2°C from pre-industrial times by the year 2100. This is disregarding the potential of a new Grand Solar Minimum.

If the temperature outcome of a massive transformation of the world's economy to reduce CO_2 emissions is the same as business-as-usual CO_2 emissions, it would be prudent to divert the money that would have been wasted trying to reduce CO_2 emissions and spend it on building the infrastructure that we will need for a 1.5°C warmer world.

We probably have at least two generations to decide whether CO_2 is even a relevant issue, and the increases in CO_2 that would have the most impact on the greenhouse gas effect are already behind us. The lack of any significant global warming in the decade and a half before and the additional years after the last IPCC report should be a wake-up call for all of us. The current popular declarations of a climate emergency are a political reaction to deflect attention from the credibility crisis is at the IPCC. Their climate forecast models are consistently wrong, and global warming/climate change/climate emergency is now unremarkable if not almost undetectable. I'll draw upon the climate heretic scientific literature to offer suggestions of where the errors may lie:

- The global temperature sensitivity to CO_2 is much lower than the educated guesses made by the IPCC.
- An assumption was made that increased CO_2 temperature increases would be amplified by more water being evaporated, and this water vapour would add to the greenhouse gas effect. This additional water vapour, with its latent heat of evaporation, was quite possibly transported by convection to the upper atmosphere where it radiated out to space.
- The main natural warming mechanism that was ignored in the models was solar activity. The sun was significantly more active in the middle of the 20th century than at the beginning of that century, or as of this writing. The IPCC acknowledges that there are changes in the sun's energy output but asserts they have little effect on the climate. They also recognize that cosmic rays may affect cloud formation, but admit that modelling clouds is problematic.

These potential errors in the models are examples of what is called confirmation bias. Assumptions had to be made, and the simulation authors made the assumptions that would push the model results to confirm what the IPCC bias was. It's not good science, but it is good politics.

Concentrating on politics instead of science is why Margaret Thatcher turned against the IPCC, which she helped found.

The Crisis in the IPCC

As Prime Minister of the United Kingdom, Margaret Thatcher had serious troubles with the one-million strong coal miners' union. They had a major strike under her watch in 1984-85 that was economically problematic enough, but Thatcher also knew that the coal itself was a domestic health hazard due to air pollution. Plus, there was already a growing concern in the scientific community that carbon dioxide, as a greenhouse gas and a significant emission from coal, was contributing to global warming. Thatcher's support of this concept was genuine, but it was also convenient that by demonizing coal and transitioning to more nuclear and North Sea oil and gas, the air quality issue would be addressed, a new domestic oil and gas industry would be supported, and the National Union of Mineworkers would be dealt another crippling blow from which they would never recover. It is not surprising that Thatcher was one of the first international leaders to call for urgent international attention to anthropogenic global warming.

On December 6, 1988, the United Nations passed resolution 43/53, which created the Intergovernmental Panel on Climate Change. The preamble to the resolution stated, "Concern that human activities could change global climate patterns, threatening present and future generations..." and then further said, "Noting with concern that the emerging evidence indicates that continued growth in 'greenhouse' gases could produce global warming..." Even before the resolution stated the mandate of the IPCC, it seems to have dictated that the IPCC's role was to confirm that human activities are causing a growth in greenhouse gases that could produce global warming. Article 10 of the resolution states what work the IPCC is to undertake, which is a comprehensive review of which part (a) is critical: "The state of the knowledge of the science of climate and climatic change."

Thatcher was all-in, and in November 1990, she stated, "The danger of global warming is as yet, unseen, but real enough for us to make changes and sacrifices so that we do not live at the expense of future generations."

By 2002, she was all-out, declaring in her book, *Statecraft,* that:

> The doomsters' favourite subject today is climate change. This has a number of attractions for them. First, the science is extremely obscure, so they cannot easily be proved wrong. Second, we all have ideas about the weather: traditionally, the English, on first acquaintance, talk of little else. Third, since clearly no plan to alter climate could be considered on anything but a global scale, it provides a marvelous excuse for worldwide, supra-national socialism. All this suggests a degree of calculation. Yet, perhaps that is to miss half the point. Rather, as it was said of Hamlet that there was method in his madness, so one feels that in the case of some of the gloomier alarmists, there is a large amount of madness in their method.

What changed her mind 180 degrees in 12 years? Now that the UK was off coal, she could admit that only 3.3% of global CO_2 emissions are from human industrial activities and the bulk of CO_2 comes from ocean degassing (57%) and animal respiration (38%), not including an unknown amount from subsea volcanoes (it is estimated that 85% of the world's active volcanoes are subsea and they emit unknown amounts of CO_2). Like other interesting lawyers in this book, Thatcher had another field of excellence; she had a degree from Oxford in chemistry and briefly worked as a research chemist. She knew the scientific method, and I suggest that's why she suspected a large amount of madness in the IPCC's methods. Let's try and reconstruct her train of thought.

The IPCC does not conduct its own research; it invites submissions from other researchers and then decides what to include in its reports. The first problem with this approach is that the UN Resolution 43/53 creating the IPCC is obviously designed only to invite research that supports anthropogenic CO2 global warming. If the IPCC reported back that a review of the state of knowledge resulted in determining climate change is natural, there would be no further need for the IPCC.

The second problem is a conflict of interest issue wherein the IPCC places the burden of proof of scientific validity back onto the very same small scientific community that was preselected because it already agrees with the IPCC mandate. It quickly becomes a mutual admiration society.

The third problem is that most impactful part of the report, the Summary for Policy Makers, is co-authored by UN member government representatives assigned to the IPCC, who did not conduct the research but who would be prone to politically influence the Summary in a way that agrees with the politicians who appointed them.

The sum of these three problems results in a takeover of climate change science by political motivations. In domestic politics, specifically in rich countries, the environmental movement is a strong, well-funded, and politically capable movement. They have been hugely successful in linking fossil fuels to environmental degradation. Many politicians, even pro-fossil fuel ones, don't want to campaign on what would be attacked as an anti-environmental platform of climate change denial. Instead they appear to be onside by kicking the issue to the United Nations with supportive gestures where any substantive action will likely occur beyond their mandate. After all, global human emissions into a shared atmosphere are an international issue. The United Nations is mostly made up of underdeveloped countries, and this may be the only forum where they have in aggregate a significant voice. A global climate change agreement that strengthens the clout of the UN and is paid for by the wealthy countries, and has tangible benefits for the underdeveloped world is good for developing nations.

The IPCC finds itself at the eye of a perfect political storm. Major democratic countries want the IPCC to deal with climate change either because they agree with the mandate or because they don't want to deal with it themselves while appearing to do so. The UN itself wants to promote more global cooperation while increasing its stature, and the underdeveloped world wants the problem, if there is a problem, to be fixed by the rich while simultaneously attempting to narrow the gap between rich and developing countries. And if the IPCC staff releases a finding that climate change is unrelated to human activity, then they may be risking ending a prestigious career posting.

The political interference in what should be a scientific inquiry goes much deeper than just the Summary for Policy Makers; it drives the direction of the scientific work itself. Research groups that produce climate change forecasts supportive of the IPCC mandate, act under the principle of publish or perish, and are more likely to be included in the IPCC reports.

Let's continue following Thatcher's change of heart toward the IPCC. The IPCC was formed in late 1988, and in July of 1990 they published the First Assessment Report, in which appeared these graphs [30]:

---

30   IPCC. First Assessment Report WG1, Climate Change the IPCC Scientific Assessment. 1990, p 202 -204.

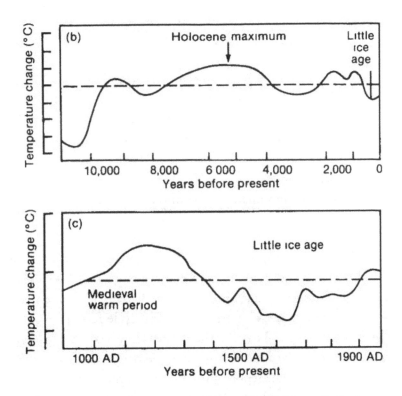

Figure #10: IPCC in 1990; Nothing to See Here, Folks!

The report's discussion of the graphs contained the following three comments that would be uncontested by academia of the day, and our Mr. Dryas would approve:

> There is growing evidence that worldwide temperatures were higher than at present during the mid-Holocene (especially 5 000-6 000 BP), at least in summer, though carbon dioxide levels appear to have been quite similar to those of the pre-industrial era at this time.

> The Early and Middle Holocene was characterized by a relatively warm climate with summer

93

temperatures in high northern latitudes about 3 - 4°C above modern values.

The Little Ice Age came to an end only in the nineteenth century. Thus, some of the global warming since 1850 could be a recovery from the Little Ice Age rather than a direct result of human activities. So it is important to recognize that natural variations of climate are appreciable and will modulate any future changes induced by man.

The IPCC's first report in 1990 supported a view that the world has been significantly hotter in the previous 10,000 years without a change in carbon dioxide, and that the then-current warming trend could be part of a natural cycle recovering from the Little Ice Age. The current temperature was very average compared to the last 1,000 years, and the temperature seemed to change naturally. The report appeared to have more of a view from standing on the shoulders of giants but left the door open to the possibility of a combination of natural and human-caused global warming in the future.

Since there is no apocalypse story here, you may wonder if this first IPCC report completed all the tasks required by the UN mandate and, thus, also be the last IPCC report. It has since been removed from the IPCC website.

This non-crisis attitude changed when, in 1999, the IPCC reviewed a submitted climate study [31], which was revised from an earlier study in 1998 by the same authors. It showed that for 1000 years before the present, the temperatures of the northern hemispheres were stable and that there was a dramatic temperature

---

31   Mann, M; Bradley, R; Hughes, M. 1999. *Northern Hemisphere Temperatures During the Past Millennium: Inferences, Uncertainties and Limitations.*

increase beginning early in the 20th century and continuing upward. Its main conclusion was that in North America, 1998 was the hottest year of the millennium. The graph of this forecast looked like a hockey stick, and the name stuck. It was the affirmation of human-emitted CO2 causing global warming that the UN was looking for in its mandate forming the IPCC, and they ran with it. Based on this single scientific research paper, which boldly eliminated the known Medieval Warm Period and the Little Ice Age while also establishing a correlation between rising temperatures and rising atmospheric CO2, the hockey stick graph appeared six times in the IPCC 2001 Third Assessment Report.

A problem quickly arose, as the hockey stick study conflicted with hundreds of previously published studies about the Medieval Warm Period and the Little Ice Age, whose temperature variations were flattened by the handle of the newly published hockey stick study. This caught the attention of politicians on both sides of the Atlantic Ocean. In testimony in 2005 to the UK House of Lords Select Committee on Economic Affairs, R. McKitrick stated, "The flawed computer program can even pull out hockey stick shapes from lists of trendless random numbers." [32]

The lead author of the is hockey stick study refused to release all his data and calculations as normally done for peers to review, but because his research used US government funds, the US House Energy and Commerce Committee was able to force an investigation into the statistical methods the hockey stick study used. The team of statisticians appointed in 2006 by the United States House Energy and Commerce Committee concluded the hockey stick study "...misused certain statistical methods in their studies, which inappropriately produce 'hockey stick' shapes

---

[32]   Plimner, I: *Heaven and Earth*, p 98.

in the temperature history." [33] This was confirmed by another committee at the US National Research Council.

The IPCC did much more than just promote a northern hemisphere temperature history that was false. In their 2001 Assessment Report a version of a second researcher's previously published work [34] was altered [35] to give the appearance of an independent verification that the research resulting in the hockey stick was correct. This was important to the IPCC as no other climate researcher had come up with results for the last 1000 years that were similar to the hockey stick.

The hockey stick study used temperatures predominately derived from tree rings, but used other proxy temperature indicators also. From the year 1000 to about 1400, the proxy data were sparse and relied more heavily on tree rings. Then from 1400 to 1900, additional non-thermometer data were available, as were tree rings. Finally, from 1900 up to the date of the publication, both tree ring data (to as late as they were available) and modern thermometer data were used. From 1900 onward, the tree ring data closely matched the thermometer data, which would give confidence that the tree ring data from 1000 to 1400 were relatively reliable indicators of temperature.

The tree ring temperatures in the hockey stick concluded that in North America from the year 1000 to 1900, the Medieval Warm Period was not as warm as today, and the Little Ice Age was not as cold as other scholars suggested. These data became the handle of the hockey stick graph. The handle showed a relatively flat and gradual cooling trend of about 0.25°C over 900 years. Then from 1900 to 1998 the study used both tree rings and thermometers, and this became the upward-pointing blade of the hockey stick. It indicated a rapid 0.75°C increase in temperatures

---

33   Plimner, I: *Heaven and Earth*, p 96.

34   Briffa, K. 2000. *Annual climate variability in the Holocene: Interpreting the message of ancient trees.*

35   www.climatediscussionnexus.com/video/Hide-the-Decline.

over this period, culminating in 1998 as the hottest year of the millennium.

The second study the IPCC used was also initially published in 1998 and updated in 2000. It studied temperatures from 1400 to 1994 using only temperature evidence from tree ring data. The second study could be easily interpreted as supporting the relatively flat handle of the hockey stick from 1400 to 1960. However, in 1961 the second study showed the temperature cooling, not warming as the hockey stick study indicated. At the same time, the thermometer data proved temperatures were warming. The main conclusion of the second study is that tree ring data may be unreliable.

The reason the tree ring data in the hockey stick study showed temperatures going up after 1960 and the tree rings of the second study showed temperatures going down is that different types of trees were used in each study. The hockey stick study used the bristlecone pine, which grows at high altitudes in California at the limit of the tree line. The bristlecone pine is highly susceptible to increases in plant fertilization, more so than temperature. It showed increased growth due to higher CO_2 levels, not higher temperatures. The second study, which used tree rings from all over North America, showed decreased temperatures after 1960 when actual temperatures were rising.

The second study highlighted that tree rings, in general, are not a reliable temperature indicator taken on their own, while the hockey stick paper knowingly used tree rings from a species that was documented to be even less reliable. Taken together, they would have shown that the hockey stick graph dependence on tree ring data during the Medieval Warm Period was unreliable and should have been discarded as evidence that 1998 was the hottest year of the millennium.

My wife and I climbed Telescope Peak in Death Valley and camped overnight among the bristlecone pine. They are like beautiful living sculptures twisted by the high winds and clinging to life literally on the edge of the rock in the dry, cold air. We could see that any environmental change for the better would help them, such as a little more CO2. They are great photographic subjects, but apparently not reliable thermometers.

In 2009, a leak of hacked IPCC emails, which became known as Climategate, revealed the efforts made in 1999 to hide the 1961 to 1994 decline in temperature of the second study that contradicted the increase in temperature of the hockey stick. To eliminate this contradiction, the IPCC used the data from the second study from 1400 to 1960, eliminated the data from 1961 to 1994, and then added actual thermometer data from 1961 to 1994. An IPCC graph of the hockey stick and the second study now appeared to be in agreement, i.e., there was a rapid warming trend after 1961, which coincided with a rapid increase in atmospheric CO2.

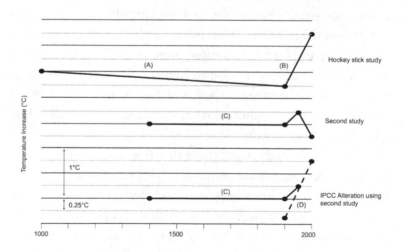

Figure #11: If the Data Don't Fit, Hide it!

The upper drawing is a simplified version of the hockey stick diagram. Segment (A) represents non-thermometer temperature data, with the bristlecone pine tree rings used predominately from 1000 to 1400, and other data added in from 1400 to 1900. Segment (B) uses both bristlecone pine tree rings and thermometers, which are in agreement. The implication was that since the bristlecone pine tree rings in Segment (B) agreed with modern thermometers, then Segment (A) is also reliable.

The middle drawing is a simplified version of the second study, which used multi-species tree ring data. Up to about 1960, this second study agreed with the hockey stick study, but the last 35 years did not. The implication was that seeing as the last 35 years did not match actual thermometer data, tree ring data were less reliable for deducing temperature and other environmental factors must be in play.

The lower drawing is a simplified version of how the IPCC presented the second study. The tree ring data after 1960 that were in disagreement with the thermometer data were dropped, and this truncation of contradictory data was hard to spot due to an overlay of thermometer data, Segment (D).

The IPCC did not disclose this alteration. It is not clear in the IPCC publications that an elimination of data had taken place, and the result was that the second researcher's work appeared to be an independent verification that the hockey stick was correct. Apparently, this was part of the "review of the state of the knowledge of the science of climate and climatic change." This was not done by any rogue scientist at the IPCC. The hacked emails detailed that the highest levels of authorship at the IPCC did not want the message of global warming diluted with contradictory views and that it wanted the reports "tarted up."

Had the original second study been used in its entirety, reviewers would have noted that both studies used tree ring data and in the second study during the late 20th century the tree ring data did not match thermometer data. Reviewers would have then questioned the reliability of tree ring data used in the hockey stick study to reconstruct temperatures 1000 years ago until 1900. As many other researchers using methods other than tree rings contend, and as the IPCC acknowledged in 1990, the Medieval Warm Period was warmer than today and the Little Ice Age was cooler than today. Such a contention supports that natural climate change is a bigger factor than fits into the IPCC's political mandate. This would eliminate the global warmers' claim that we are living in the warmest period in the last 1000 years, and cast doubt that greenhouse gases are largely responsible for recent warming.

It also brings into question the integrity of the IPCC and the need for an independent audit of their work. Consider the chronology of events:

- In 1988, the IPCC is founded with the support of Margaret Thatcher.
- In 1990, the inaugural IPCC report endorses the conventional scientific view that the Medieval Warm Period and the Little Ice Age existed in the last millennium, and that neither was caused by CO2 or human activities. Current climate change appeared to be historically normal, and current temperatures are cooler than in medieval times.
- Then in 1998, the hockey stick graph is published as an independent scientific paper covering the years from 1400 to date, then revised and republished in 1999 to include the years from 1000 to date.

- In 1999, the IPCC conspires to alter a second study similar to the hockey stick, resulting in the appearance that the hockey stick study is independently verifiable.
- The IPCC publishes a significant report in 2001 with the hockey stick featured six times. The altered second study is included to support the legitimacy of the hockey stick graph. The Medieval Warm Period and the Little Ice Age are missing, and 1998 is claimed to be the hottest year of the millennium.
- Margaret Thatcher publishes *Statecraft* in 2002, wherein she says about climate alarmists, "...there is a large amount of madness in their method."
- The United States House Energy and Commerce Committee in 2006 determined the hockey stick graph was a result of misused statistics.
- In 2006, the movie *An Inconvenient Truth* is released, prominently featuring the hockey stick.
- The IPCC, once again, uses the hockey stick graph in a 2007 report.
- In 2009, Climategate occurs, and the hockey stick disappears from future IPCC publications.

In most professions, including my own, a breach of the public trust like the hockey stick graph would raise questions about gross negligence and willful misconduct. Furthermore, if real damage was done, revocation of the licence to practise, or even jail, could result. Would you ride in a helicopter because the flight simulator said everything was okay, even if the input data were changed to trendless random numbers and the results were the same? Would you "tart up" the safety results if your boss told you to?

Without the hockey stick graph in 1999, I am wondering if we would still have the Paris Agreement of 2015, and the intervening global climate hysteria. The IPCC still retains the hockey stick mentality, but without the untruthful graph and the publicity of *An Inconvenient Truth,* perhaps the public may never have bought in.

And then the whistleblowers came out. After the 2006 debacle of the hockey stick graph, the IPCC should have been motivated to ensure a high degree of confidence in the material on which it based its sweeping and world-changing recommendations. The IPCC relies heavily on a global thermometer temperature database called HadCRUT4. The name derives from marine temperature records kept by the Hadley Centre of the UK Met Office and terrestrial temperature records kept by the Climatic Research Unit of the University of East Anglia. The relationship between HadCRUT and the IPCC began with Margaret Thatcher and has continued ever since. The 2009 email hacking incident exposed the Climatic Research Unit to charges of manipulating data, dubbed Climategate, of which multiple investigations exonerated them. Scientific misconduct was ruled out, but not error.

When John McLean was an Expert Reviewer of the IPCC's fifth climate assessment report in 2013, he asked if the HadCRUT data had been audited, and the answer he received was that it was not the job of the IPCC to do so. McLean was not satisfied with that answer and effectively became an IPCC whistleblower. In 2017 McLean earned a PhD from James Cook University in Australia with his thesis extensively titled, *An audit of uncertainties in the HadCRUT4 temperature anomaly dataset plus the investigation of three other contemporary climate issues.* Remarkably, it was the first audit ever done on a data set that has influenced world treaties and hundreds of billions of dollars in expenditures. If not for the seriousness of the global stakes, the audit findings regarding quality control may have been considered humorous, with erroneous claims of a Colombian town experiencing an average temperature over 80°C in 1978 for three months, and marine temperatures logged at locations 100 km inland. These items would tend to be averaged out if they occurred at random. However, it was three major systematic flaws in the HadCRUT4 database, uncovered by McLean, that are the most damaging:

- The data set begins in 1850, and the thermometer records are heavily biased by European countries, their colonies, and their trade routes. In 1850 there were temperature data for only 20% of the Earth's surface. The temperature data grew in roughly 20% increments every 50 years until, in the year 2000, there was temperature data coverage for 80% of the globe. When the IPCC claims a global temperature increase since 1900 and quotes the HadCRUT4 database, those data are based on thermometer measurements in the year 1900, which cover only about 40% of the Earth, and are then extrapolated to the rest. It assumes the post-1900 changes to temperature are homogeneous around the globe. In essence, the global average temperature calculated in 1900 is heavily weighted to Europe.

- While oceans cover 71% of the Earth, the subsurface sea temperature data make up 82% of the HadCRUT4 database. This is due to ships in port recording water temperatures in the captains' logs that were also used as part of the land temperature database. Prior to 1950 there was no standardized method, or depth, to take the temperature of the oceans just below the surface. As both the method of taking the sample and the depth of the sample can affect the temperature reading dramatically, McLean advises that pre-1950 subsurface sea temperature data are unusable. For the period from 1900 to 1950, approximately 50% of the Earth's surface had thermometer temperature recordings, and 82% of that coverage was from unreliable subsurface sea measurements. The math nerds will have calculated that less than 10% of the data used by HadCRUT4 are from reliable physical temperature recordings between 1900 and 1950.

- Post 1950, the marine data are much better, but McLean still has reservations about how the entire data set was statistically used. HadCRUT failed to follow important World Meteorological Organization protocol on minimum data requirements, which increases the risk of inaccurate global average temperature calculations.

In summary, the former IPCC Expert Reviewer determined in a PhD thesis that the global temperature database preferred by the IPCC was grossly deficient prior to 1950, was not screened for data input errors, and failed some prescribed statistical standards. The global warming community likes to use this database to further their claim that the global average temperature has warmed up by 0.8°C since 1880.

In 2015, whistleblowers [36] at the USA National Oceanic and Atmospheric Administration said their agency doctored global numbers to hide evidence that the Earth's temperature had hit a plateau in 1998 (see Figure #9), and that this plateau was hidden from world leaders to promote the signing of the 2015 Paris Agreement. A Danish delegation was refused by the IPCC to submit a peer-reviewed research paper already prestigiously published that showed the link between global warming and solar activity, and that cosmic rays seeded clouds (as explained in Chapter 2). [37] An ex-IPCC author, who contributed to sea level measurements, resigned and publicly claimed that the IPCC misused his work. [38] Fifteen co-authors of the 2013 IPCC report published a separate paper (with the addition of two non-IPCC authors) stating they acknowledge the 1998 global warming "pause." They also acknowledged a lesser sensitivity of temperature increase to CO_2 increase than shown in current climate model

---

36 www.dailymail.co.uk/sciencetech/article4192182.
37 Solomon, L. 02Sept2011 Financial Post.
38 Newman, A. 12Feb2019 The New American.

simulations. [39] The list of whistleblowers is very concerning. It is not my intent to address the grievances on that list but rather to point out that the scientists who disagree with the IPCC include many from within the IPCC community itself.

In any other sector, whether public or private, this kind of performance would make heads roll. In the IPCC world there appears to have been no consequences, which is understandable in a large global political bureaucracy with tremendous clout and no accountability. The global warming community simply cut off the historical shaft of the hockey stick that was problematic and continued with the forecast of the blade. They continued to endorse computer global warming simulators, which have become more imminently apocalyptic but took great care to achieve a better, but shorter, history match. In confirmation of the pre-ordained bias of the UN Resolution 43/53, the IPCC has accepted only these simulations, averaged them into an overall forecast, and published them as "the state of the knowledge of the science of climate and climatic change."

In the Introduction I stated that, as a Professional Engineer, I was motivated to write this book because of the omission, alteration and fabrication of evidence that is being used to support the notion of a climate emergency:

- At the IPCC, the omissions of evidence are: not calculating the solar effects on climate change; the near silence that CO_2 as a greenhouse gas has a practical upper limit (the logarithmic effect); and minimal discussion that water vapour makes up 90% of the greenhouse gas effect.
- The alteration of evidence is the hockey stick graph initially wiping out known global warming and cooling periods, and then when that failed, shortening global

---

39   Moran, A. et al. 2015: Climate Change The Facts, p 62.

temperature history to the end of the Little Ice Age (when the Earth was warming naturally).

- The fabrication of evidence is promoting the computer simulation global temperature forecasts as "the state of the knowledge of the science of climate and climatic change," when in fact, the IPCC reports are not science; they are documents to support the political consensus of the United Nations.

This all adds up to a credibility issue at the IPCC. And there is more: their climate forecast models don't work; they attempted to obscure the 1998 to 2012 global warming pause; they have a queue of whistle-blowers; and they have failed utterly in providing "the state of the knowledge of the science of climate and climatic change."

There is no climate emergency. The IPCC cannot invent one, and so, they are faced with a crisis. Politics in equals politics out.

What then, is the legacy of the IPCC? Again, I will defer to Margaret Thatcher with her views of the 1997 Kyoto Protocol of the United Nations Framework Convention on Climate Change: "Kyoto was an anti-growth, anti-capitalist, anti-American project which no American leader alert to his country's national interests could have supported." We will see in a later chapter that the 2015 Paris Agreement might have garnered the same reaction from her.

CHAPTER 5

The Good, The Bad, and The Ugliest CO2 Reduction Ideas

In the last few decades there have been many large-scale efforts to reduce human emissions of carbon dioxide, and many of them have involved significant subsidies by the taxpayer. They are mostly justified by the political viewpoint of the Big Picture; in essence, the subsidies are a small price to pay to get us off the hydrocarbon addiction and on to a green energy future. Let's examine some of these initiatives, but this time from a Professional Engineer's viewpoint of the Big Picture. Rather than looking at only CO2 reduction, we need to consider the overall impact these five common carbon dioxide reduction initiatives have on the environment and our quality of life.

Good CO2 Reduction Ideas

Banning Coal

One of my favourite lines from a politician is from one who, when he got elected, was cornered by an unrelenting reporter into admitting his predecessor had made the right call on a big decision, a call that the victorious politician had campaigned against. His post-election response was, "Yes, it is obvious he made the right decision, but I have always said he made the right decision for the wrong reasons."

Banning some types of coal is a good environmental decision, but not because it reduces CO_2 emissions. Coal has two types of harmful emissions that are hazardous to human health. The first is particulate matter, particles less than 10 micrometres in diameter (you may recall that an average human hair thickness is 10 times larger at 100 micrometres). These particles stay suspended in the air and are classified as aerosols (solid particles floating in the air). These particles are just the right size to be inhaled by humans and cause a multitude of problems in our lungs. Medical research is pointing toward particulate matter less than 2.5 micrometres in diameter as the most harmful. The *Global Burden of Diseases Study 2015*, in part funded by the Bill and Melinda Gates Foundation, found that 7.6% of all deaths globally (over four million deaths) are attributable to exposure to 2.5 micrometre particulate matter in the air. 59% of these deaths are in east and south Asia. [40]

The second coal emission that is harmful to humans and animals is other chemicals contained in some coals as impurities. The worst case scenario to date was a five-day coal induced smog over London in 1952 that killed about 12,000 people. It was a perfect storm of atmospheric chemistry. When coal is burned it produces nitrogen dioxide, and because the coal being burned in the UK at the time contained sulphur, another emission was sulphur dioxide. Typically, this would result in ordinary sulphurous smog, but during those five days in 1952, London was experiencing a natural fog also. The combination of nitrogen dioxide, sulphur dioxide and water droplets resulted in sulphuric acid, which Londoners inhaled. In addition to the deaths, 150,000 people were hospitalized.

The release of chemicals with coal is not restricted to sulphur, as arsenic can also be released. Coal may contain heavy metals

---

40   Cohen, A. et al. 13 May 2017, Lancet, pages 1907-1918

like mercury, chromium or even lead, and when we mine that coal and burn it, those heavy metals are released into the atmosphere. 26% of the global mercury emissions are from the burning of coal. [41] The purest coal to burn is Anthracite as it has the fewest impurities and highest heat content, followed by Bituminous, Sub-bituminous, and Lignite, which is the most impure form of coal.

The burning of dirty coal needs to be restricted not because of the CO_2 released, but because of the health hazards. Those most susceptible to harm include children, the elderly, pregnant women, and people with lung conditions. There has been much medical research on the health hazards of coal, but a notable non-health hazard is carbon dioxide. Air that was unfit to breathe was the reason the United Kingdom, especially London, had to get off the coal habit in the 1950s. History is being repeated in Asia right now. Banning coal is a good idea. The release of CO_2 won't harm you, but particulate matter, sulphuric acid, mercury and lead will. The onus should be on the coal industry to filter out the particulate matter and remove the impurities if they are to be allowed to continue burning coal.

Promoting Hydroelectricity

Hydroelectricity is a great energy source with several ancillary benefits. Many dams and reservoirs are more than a sustainable electricity source as they often provide potable water storage for major urban areas, enhance agricultural opportunities, assist in flood control or mitigation, and create new waterfront and recreational property. There are well-documented costs associated with all these benefits: people (often Indigenous) and wildlife are displaced; terrestrial habitat is changed to aquatic habitat; and downstream habitats have diminished water supply. For global warmers, these negative environmental impacts are more than

---

[41]   Burt, E. Et al. April 2013; *Scientific Evidence of Health Effects from Coal Use in Energy Generation;* University of Illinois at Chicago.

offset by the generation of electricity without CO_2 emissions, because no hydrocarbon fuel combustion is involved.

What is often missed is that hydroelectric reservoirs give off a more powerful greenhouse gas—methane.

On the western edge of the Great Plains in Canada, the Bighorn Dam was built in 1972 to generate electricity. The reservoir is called Abraham Lake, and it covers a shale and limestone basin in the Rocky Mountains, which grows pine trees surrounding the man-made lake. Each spring the mountain snowmelt is captured in Abraham Lake, power is generated all summer, and by fall the lake level is very much reduced and has the capacity for another cycle of snowmelt in the next spring. Every winter thousands of tourists, many of whom are photographers, descend on the tiny nearby village of Nordegg to visit the frozen lake and take pictures of the frozen methane gas bubbles trapped in the ice. I have visited the lake. Some of the trapped bubbles have created ice-hills three metres high, and the bubble fields cover many hectares.

There are no oil or natural gas deposits or production near Abraham Lake. After 48 years of submersion, any previous vegetation should have long been rotted away. The source of the methane is decaying plants growing on the bottom of the lake, plants that are nowhere visible on the surrounding terrain. Rotting organic material creates what we term biogenic methane, and this gas builds up on the bottom water layer of the lake. When the lake level is lowered in the late fall, the resulting loss of hydrostatic pressure on the methane causes it to expand and bubble up to the surface, which has already formed a thin layer of ice. Then the ice freezes to the bottom and traps the bubbles, and the village of Nordegg gets to host winter-hardy tourists seeking a unique natural phenomenon that is actually man-induced.

#12: Frozen Greenhouse Gas (Methane) Bubbles in Abraham Lake

Hydropower isn't greenhouse gas emissions-free, nor does it have zero environmental impact.

The Bighorn Dam and Abraham Lake have created a new and abundant methane emission source. Depending on where you source the information, methane has a heat-trapping capacity as a greenhouse gas of anywhere from 20 times to 84 times more than CO2 emissions. Abraham Lake is not unique. A detailed sampling of 10 hydroelectric reservoirs in Brazil estimated that compared to combined-cycle natural gas electrical generating plants (which are the most efficient design), the 10 Brazilian dams in aggregate put out up to 18.2% of the equivalent greenhouse gas emissions for the same amount of electricity generated.[42] Each hydroelectric reservoir in the world likely releases a different amount of methane due to the amount of organic material washing into or contained in the water, the temperature of the water, the age of the reservoir, and possibly other factors not yet understood. An effort to determine these variables has been coordinated by the

---

[42]   Dos Santos, M.A., Rosa L.P. 2011; *Greenhouse Gas Emissions from Hydropower Reservoirs: A Synthesis of Knowledge.* Hydropower and Dams, Issue 4.

United Nations Educational, Scientific and Cultural Organization (UNESCO), who conducted a research project from 2006 to 2012 to estimate the greenhouse gas emissions from freshwater hydroelectric reservoirs. As a result, the latest guidelines issued by the IPCC [43] in 2019 now have a methodology for calculating both carbon dioxide and methane emissions from both new reservoirs (which emit more due to submerged pre-existing biomass) and older reservoirs.

If the objective is to reduce greenhouse gas emissions, then any step away from coal-fired power stations should be a step forward. Natural gas produces fewer greenhouse gases than coal, and hydropower produces fewer greenhouse gases than natural gas. And both emit almost no dangerous particulate matter or heavy metals that are prevalent in coal. If the objective is to do what is best for the environment, hydropower could in fact be worse than natural gas generated power. First, factor in the significant environmental impact caused by the footprint of the hydroelectric dam and reservoir, and the amount of CO_2 given off by the cement used. Then consider that burning natural gas emits mostly carbon dioxide and hydroelectric dams emit mostly methane. If the concentration of carbon dioxide in the atmosphere has already passed the point where more CO_2 has little effect (the logarithmic relationship again), then more CO_2 would not add to the greenhouse gas effect, while methane still would. In that scenario, hydroelectricity could be both a much larger footprint of ecological disturbance and a greater contributor to the greenhouse gas effect than natural gas-generated electricity.

The next IPCC Assessment Report, AR6, is scheduled for 2022, and the Physical Science Basis for that assessment is due in 2021. Based on the new 2019 guidelines, they should estimate how large a greenhouse gas contribution hydropower makes, and

---

43   IPCC. May 2019; 2019 Refinement to the 2006 IPCC Guidelines on National Greenhouse Gas Emissions; Volume 4, Chapter 7.

perhaps then the free ride of hydropower as a non-emitter of greenhouse gas may end.

Intrepid, frostbitten photographers of the Bighorn Dam and Abraham Lake have known for a long time that hydropower reservoirs emit methane. My wife is one of them. To give the IPCC credit, they appear to acknowledge that now. It will be interesting to see how governments who have relied on their hydropower to meet greenhouse gas emission targets react to the new reality that hydropower emits methane. It will get very interesting if the IPCC shifts its greenhouse gas focus from CO2 to methane once its politicians catch up with its scientists regarding the logarithmic effect of CO2.

Hydropower is still a good idea, but it is not environmentally consequence-free.

Bad CO2 Reduction Ideas

Windmills and Solar Panels

There are numerous environmental complaints about windmills and solar panels that don't seem to apply (at least in the media) to traditional electrical power generation sources. With windmills, the main environmental complaint is bird kills. The Audubon Society published an estimate of bird kills that worked out to approximately five birds per wind turbine per year [44], which with roughly 50,000 wind turbines in the continental USA would work out to 250,000 birds per year. While this is a significant number, many argue it is still small compared to the total number of birds killed by human activity, estimated by the US Fish and Wildlife Service to be about 3.3 billion per year. [45] From a fairness perspective, the limited press coverage of the ongoing annual wind turbine bird kills should be compared to the international press

---

44 Bryce E. 16 March 2016, Audubon.org. *Will Wind Turbines Ever Be Safe for Birds?*

45 Fws.gov/birds-enthusiasts/threats-to-birds

coverage of a one-time incident of 1600 ducks killed on an oil sand mining site in Canada in 2008. The oil sand operator was fined $3 million for failing to have in operation existing devices that would have prevented the deaths. When the Audubon USA kill rates are extrapolated to Canada, Canadian wind turbines probably kill at least 35,000 birds per year, yet there are no fines and no required wind turbine impact prevention devices. It is discriminatory to hold the hydrocarbon energy industry to a much higher standard than the wind turbine energy industry.

As for solar panels, most of the negative environmental issues have been resolved. An original issue was that for the first-generation photovoltaic cells, they took more energy to manufacture than the cells would ever produce. However, this has now been resolved, and the third-generation photovoltaic cells produce several multiples of energy more than it took to make them. Many solar installations are coupled with a lithium-ion battery for electric storage, and the extraction of lithium, as in any mineral extraction operation, can be very damaging to the environment if not regulated and policed properly.

What links wind and solar power together in the classification of Bad Ideas, is that they don't always work because both power supplies are dependent upon either the wind blowing or the sun shining. If you were the engineer in charge of supplying your community's electrical grid from wind and solar, here are the three biggest headaches you would have:

- Variability of supply–When the sun is shining and the wind is blowing, you cannot control the wind speed or the cloudiness, so the amount of power available is not within your control. Yet, you must accept into the grid whatever wind and solar power is available.
- Uncertainty of supply–Weather forecasts for wind and sunshine are not reliable long term or seasonally, so it is

difficult to plan even a day or two in advance without knowing how much wind and solar power will be available.

- Load factor–If your community is in the USA, wind power will produce about 34% of its designed output capacity annually, and solar will be about 28%. In the long term, you cannot rely on wind or solar to keep the lights on for your customers.

The single answer to all three issues is to have a 100% demand coverage backup conventional fuel electricity generation plant. And in that scenario, having wind and/or solar electrical generation backed up by a conventional hydrocarbon fuel electrical generation plant means the capital invested in electrical power generation is higher than need be by exactly the cost of the wind and solar plants. To recoup that sizeable capital investment, either the power utility must raise electricity rates, or the taxpayer must subsidize the electricity bills, or both. If you are adamant that wind and solar are necessary to reduce CO_2 emissions, then to reduce the cost of the backup conventional power plant, the owner or government is incentivized to make that backup power plant run on the cheapest fuel available—coal. Wind and solar power incentivize the use of coal as a needed backup power supply.

That very scenario happened in Germany. They spent hundreds of billions of dollars on solar and wind power, had trouble manufacturing BMWs by candlelight (my exaggeration), and are now deferring the shutdown of coal-fired power plants and expanding the domestic mining of CO_2 emitting lignite, the most impure form of coal. Why this failure is significant to green energy proponents is that Germany was supposed to be the inspirational champion for the transition to renewable green energy. Germany has the fourth-largest fleet of coal-fueled power generation plants in the world. They produce 40% of Germany's electricity, and they alone produce 7% of the European Union's CO_2 emissions. It is also a very wealthy nation, renowned for its engineering excellence, and it embraced solar and wind power with over $500 billion in investments. If the German nation could not transition from hydrocarbon fuels to green power, who could? They are now falling back on coal to backstop solar and wind and now have the highest electricity prices in the European Union, almost 60% higher than the UK, despite having to give away power for free when the wind blows too hard. Had they phased out coal, skipped the wind and solar installations, and phased in natural gas-fired power plants, they probably would be ahead both economically and on the CO_2 scorecard.

Wind and solar power are good ideas when they operate in isolation and are designed for a specific, dedicated use, but they are bad ideas for replacing an established national utility grid.

Switching from Gasoline to Diesel

I could not write it better than the Mayor of London, Sadiq Khan [46], when he stated, "The problem is that governments often fail to grasp that focusing on one issue at a time, such as CO_2 output, inevitably leads them to ignore others, such as toxic emissions." After a significant effort and expense to promote diesel engines in the UK, his conclusion was: "The science now tells us that diesel vehicles cause more than four times the pollution than petrol cars."

In an effort to reduce CO_2 emissions under the Kyoto Agreement, UK governments promoted a switch from petrol (gasoline) to diesel, which has less tailpipe CO_2 emissions. The pollution Mr. Khan is referring to consists of the significantly increased nitrogen oxides and particulate matter released by diesel compared to gasoline. A review of the available research

---

[46]   Khan, S. 22 May 2017. *Fact Check: are diesel cars really more polluting than petrol cars?* www.Theconversation.com

[47] indicates that particulate matter of the 2.5 micrometre size in diesel exhaust may be more harmful because more of it is made up of black carbon. The real problem, as Mayor Khan points out, is that switching from gasoline to diesel cars increases the amount of nitrogen oxides emitted about 45%. Nitrogen oxides are not good for human health, but one of them, nitrogen dioxide, is especially aggravating for respiratory conditions, specifically asthma, a condition which the mayor has recently developed.

Nitrogen oxides are the chemicals that react with sunlight to make yellow-tinged photochemical smog, as found in Los Angeles. In 60 years, London went from having unhealthy air due to sulphurous smog caused by sulphur dioxide (known as London Smog) to having unhealthy air due to photochemical smog caused by nitrogen dioxide (known as Los Angeles Smog). The correct answer would have been electric vehicles and no smog. This is what is behind the Chinese program to promote electric vehicles; it is not to reduce carbon dioxide emissions, which would still be produced at the coal-fired electricity generation plant, but to minimize the Los Angeles Smog over Beijing.

The Ugliest CO2 Reduction Idea:
Food Converted to Fuel

Biofuels are as old as the harnessing of fire. At first, anything with a cellulose component would do (wood, straw, dung., etc.). However, the energy released has always been inferior compared to the high-energy density and storage convenience of coal, whale oil, or crude oil. More recently, agricultural and food industry waste products or surplus agricultural products were usefully converted to energy products such as biodiesel or ethanol in times of high crude oil prices or global crude oil supply insecurity. It was the

---

47   Hime, N.J. et al. June 2018; *A Comparison of the Health Effects of Ambient Particulate Matter Air Pollution from Five Emission Sources, International Journal   of Environmental Research and Public Health,* page 1206.

next step that began a slide down the slippery slope of moral hazard. As a measure to reduce American dependence on foreign oil, President George W. Bush mandated minimum ethanol content in gasoline. The unintended consequence of this was to link American farmland use to the demand for gasoline. Growing corn for ethanol now came into direct competition with growing any crop, including corn, for food. Then we slid further down the slope as many other countries have now mandated that biofuels, such as ethanol or biodiesel, be used in transportation fuels as a way to reduce CO_2 emissions.

We have now unintentionally globalized the linkage of the world's farmland to the demand for both gasoline and diesel. The farmers of the world are now incentivized to grow crops for either fuel or food, depending on the price of biofuels. There are three main negative consequences to establishing this food or fuel link:

1. Converting crops from food to fuel will hurt the world's poor the hardest while the rich fill their fuel tanks with what was previously destined as food for the poor.

Before the US Energy Independence and Security Act of 2007, some corn-based ethanol was used in the USA to increase the octane rating in gasoline. In 2005, 14% of USA corn was converted to ethanol, and the average gallon of gasoline contained 3% ethanol. By 2012, gasoline was mandated to contain 10% ethanol, and 40% of all USA corn was converted from food to fuel. This was equivalent to 15% of all the corn grown in the world. About 3.3% of global grain production was used in the USA as a fuel additive to lessen American dependence on foreign oil. This was more corn than was consumed on the entire continent of Africa. The price increase in corn attributed to this diversion from food to fuel was 30%. [48]

Corn is the most essential staple food in the world. Globally, it makes up about 20% of all the dietary calories consumed by humans. People in rich countries spend less than 10% of their household income on food that is consumed at home, while people in developing countries spend 40% to 50% on food. It follows that an increase in corn prices would hurt the poor significantly more. A 30% price increase took place when corn was diverted from food to fuel to protect the USA from unsecured foreign oil imports, but the USA is now the largest oil producer in the world and currently exports crude oil. Some of the world followed suit and mandated ethanol in gasoline to reduce CO_2 emissions (it doesn't, more to follow) on the basis that CO_2 emissions cause global warming. While I do not begrudge farmers for having a better income, and even if oil supply and global warming were major concerns, it is particularly immoral that we passed a disproportionate cost of oil security and CO_2 reduction onto the poorest people in the world. It is unconscionable if we don't even need those programs anymore.

2. In some instances, the production of biofuels creates incentives for environmental degradation.

In addition to blending ethanol into gasoline, the USA mandated refineries in 2007 to use a minimum percentage of renewable oil feedstock to make diesel. The idea was that waste cooking oil, waste animal fats, or oils from

---

[48] *Carter et al. 2013; The Effect of the US Ethanol Mandate on Corn Prices. University of California at Davis, Department of Agricultural and Resource Economics.*

surplus agricultural crops could enhance fuel security, but again, the initiative caught on world wide as a carbon dioxide emission reduction idea. Soon, there were not enough waste products or surplus agricultural oils left to go around, so farmers grew them. This is different from the food or fuel land use argument because, in this case, previously non-agricultural lands are being converted into plantations to produce biofuel feedstock. It sounds worse when you say jungles and rainforests are being burned to grow palm oil. Palm oil turned out to be the highest yielding and highest profit renewable oil, and it grows best within 10 degrees latitude of the equator. In response to the rapid expansion of production of palm oil at the expense of tropical deforestation, the European Union recently moved to limit the purchase of biofuel feedstock produced from lands where it causes deforestation, directly or indirectly, that would not have otherwise occurred.

Let's have another look at ethanol and biodiesel from a different angle. Perhaps environmental degradation is too strong a term, but environmental stress isn't. Again, we will use USA statistics because they are readily available for both corn (grown for ethanol) and soybeans (grown for biodiesel). Irrigation for agriculture in the USA accounts for 38% of all the water withdrawn from lakes, rivers and aquifers. Corn makes up 25% of all the irrigated acreage, and soybeans account for 14% of the irrigated acreage. Since 40% of the corn winds up as ethanol, it is not a stretch to make a rough estimate that it uses about 4% of the water. Likewise, about 15% of the soybeans end up as USA domestic biofuels, resulting in another 1% of the water being used. While this is not an exact accounting, it does suggest that 5% of all the water withdrawn from lakes, rivers and aquifers in the USA is used to produce biofuels for a security of supply issue that has passed, and a CO_2 emission issue that they do not help with and may not even be relevant. What is the environmental cost of 5% of the water supply in the USA? I would doubt any homeowner on water rationing would be happy about this.

3. When adding corn-based ethanol to gasoline, there is no reduction in hydrocarbon fuels consumed, and carbon dioxide emissions are unaffected.

A debate about corn-based ethanol has been simmering in the engineering world for decades (simmering is about as exciting as we get). The question is: when accounting for the fuel used in the corn farming operation, plus the energy consumed by the ethanol production process, is more energy put into the production of ethanol blended gasoline compared to what is subsequently released in an internal combustion engine? For a long time the consensus was that there was more energy in than energy out, as it used to be with solar panels. The range of carefully considered numbers I am aware of show that over time, technical efficiencies have resulted in the fact that ethanol now contains between 25% [49] and 38% [50] more energy than it took to produce it. When 10% ethanol is blended with 90% into petroleum gasoline, the blend (sold as E10 gasoline) now has an energy gain benefit from the production of ethanol of only 2.5% to 4%. However, ethanol releases during combustion $^1/_3$ less energy than gasoline, so when you mix 90% petroleum gasoline and

---

49   *Ibid.*
50   *Lorenz D and Morris D, Aug 1995. How Much Energy Does It Take to Make a Gallon of Ethanol? Institute for Local Self Reliance (large. stanford.edu).*

10% ethanol and burn it in a car, it takes 3% to 4% [51] more of the E10 blend to travel the same distance as the 100% petroleum gasoline. Whatever energy gain realized in producing ethanol (2.5% to 4%) is lost by having to burn more (3% to 4%) of the E10 blend to go the same distance.

Another debate about ethanol added to gasoline is the question of whether the resulting combustion as an automobile fuel results in reduced carbon dioxide emissions. One would think this would be a no-brainer since most automotive fuel advertising promotes a 10% ethanol gasoline blend as a green fuel. Recall that ethanol was first introduced into gasoline to increase the octane rating (to make the gasoline burn more completely). It was then mandated to be used in the USA to promote fuel supply security but was never added to gasoline to reduce carbon dioxide emissions at the tailpipe. Ethanol is now being promoted and advertised as a carbon-neutral fuel on the basis that whatever CO_2 is released by burning it, the same amount of CO_2 was used in photosynthesis to produce the corn. So, the energy derived from burning the ethanol is considered carbon-free and is not included as a new emission to the atmosphere. Extrapolating that logic should result in justifying clear-cutting of forests for firewood to run steam locomotives, which would be CO_2 neutral by this accounting. James Watt, the inventor of the stream locomotive, was almost 250 years ahead of his time!

The reality is when you burn ethanol in gasoline, it still releases the same amount of CO_2 into the atmosphere to

---

[51]   www.fueleconomy.gov/feg/ethanol.

drive the same distance as though no ethanol was added. Furthermore, if the corn was not converted to ethanol, corn would still have been grown for food or some other crop would have been grown. There are seven billion mouths to feed on this planet, and no farmland would go idle if we stopped putting ethanol into gasoline. A similar amount of CO_2 consuming photosynthesis would still take place.

The same misguided philosophy has been applied to biodiesel. It assumes that more photosynthesis has taken place just because the plant is destined to be used as fuel. Photosynthesis from some other plants on that land would have taken place whether or not biodiesel was produced, possibly by a jungle saved from burning. Therefore, ethanol and all other biofuels should not be given a free ride on CO_2 emissions.

Here is how the math works [52]:

- Let's assume that the farmland was going to be farmed anyway, so any CO_2 emitted in the corn farming operations is not counted.
- The process of converting corn to ethanol produces about 6.6 lbs of CO_2 per US gallon of ethanol produced. Much of this CO_2 is very pure and is captured and used in the food business of other industries, so we won't count it either. However, the market for pure CO_2 is not unlimited, and ethanol production co-produces a lot of CO_2.
- One gallon of 100% petroleum gasoline releases 19.64 pounds of CO_2 at the tailpipe.

---

[52] www.eia.gov/tools/faq *how-much-CO2-is produced-by burning-gasoline-and-diesel-fuel*

- One gallon of E10 is listed for national greenhouse gas inventory purposes as releasing 17.68 pounds of CO_2 at the tailpipe. Only the petroleum-derived CO_2 is considered; the ethanol CO_2 is not.
- One gallon of E10 physically releases 18.95 pounds of CO_2 at the tailpipe, or 96% of the CO_2 emissions from 100% petroleum gasoline.
- Since it takes 3% to 4% more E10 to go the same distance as 100% petroleum gasoline, the total CO_2 released at the tailpipe is 18.95 X 1.035 = 19.61 pounds of CO_2 at the tailpipe for the same distance driven.

If the CO_2 co-produced in the ethanol production process is vented to the atmosphere, then another 0.68 lbs of CO_2 would be released, making E10 gasoline CO_2 emissions slightly higher for the same distance driven as 100% petroleum gasoline.

There is another problem with biodiesel that drives up the price of fuel. If a refinery was not designed to process biofuels (for example, it was designed to process heavy crude), and if by government mandate it must produce a diesel product that was made with 10% biofuels as feedstock, the efficiency and most likely the capacity of the refinery is degraded because it is operating under suboptimal parameters. This drop in efficiency or capacity increases the price of the fuel, and the fuel produced will physically emit about the same amount of CO_2 as before. To be fair, biofuels feedstocks are sulphur-free, so the resultant fuel may have a lower sulphur content, which is good for the environment.

Here is a summary of four reasons why biofuels, especially ethanol, take the Ugliest Spot for CO_2 reductions ideas that were not thought all the way through:

- They don't physically reduce CO2 emissions.
- Corn ethanol does not materially reduce the use of petroleum fuel when you consider the reduced mileage of the petroleum-ethanol blend gasoline.
- Biofuels incentivize tropical deforestation and use freshwater supplies for irrigation.
- The worst consequence is that ethanol from corn makes food significantly more expensive for the world's poorest.

We should not pat ourselves on the back for embracing biofuels; we should stop using them.

CHAPTER 6

Quaecumque Vera (Whatsoever Things Are True)

My wife and I hired a guide to take us kayaking around some islands near Tofino, on Vancouver Island, in Canada. The young man was well-read in environmental issues, had been doing the guiding gig for several years, and obviously loved what he did. There were just the three of us, and we were hoping to see some Orcas (Killer Whales), but it was late in the season for that. In the media, the popular viewpoint accepted as "settled science" was that global warming was threatening the very existence of the Southern Resident Killer Whale near Vancouver Island. I asked the guide if this was true, and got a big belly laugh in response. "That's all BS! The southern resident killer whale eats mostly Chinook salmon, and the Chinook salmon eats mostly herring, and we have fished out the herring. The killer whale is dying out because we overfished the base of his food chain."

His statements all checked out; the actual numbers are 82% of the killer whale diet is Chinook salmon, 62% of the Chinook salmon diet is herring, and the pacific herring fishery collapsed in the 1960s due to overfishing. Yet, virtually every news outlet was blaming global warming for the demise of the Southern Resident Killer Whale, when it was, in fact, the typical story: reckless exploitation of an ecosystem that upset the natural balance with unintended consequences. In somewhat of an unhappy ironic twist, the killer whales use echolocation to find the Chinook salmon, and the reduced stocks of Chinook salmon are harder to find because higher frequency marine engines interfere with echolocation. These higher frequencies originate from smaller engines on smaller vessels. Local global warmers want to ban oil tankers to prevent collisions with the Orcas, but the real problem is that the underwater noise from whale watching boats interferes with the Orcas' hunting. A further negative ecological factor in the ocean around Vancouver Island is that the capital city, Victoria, dumps its raw sewage into the Pacific, but the press is happy to report that global warming is killing the killer whales.

Other examples of humans causing extinctions or endangerments occurred when the First Nations immigrated across the Bering Strait and settled in North and South America some 15,000 or so years ago, and the megafauna (ultra-large mammals) of both continents became extinct. When the Europeans first started fishing in the Grand Banks off the east coast of North America, Captain John Cabot boasted that the masses of cod "slowed the progress of the ships." 500 years later, overfishing caused the collapse of the greatest fishery on Earth. After the American civil war ended 150 years ago and pioneers raced westward to settle the Great Plains, the great bison herds that had sustained the First Nations for millennia were rapidly killed off. Then 100 years ago, the tuna fishing grounds off southern California were overfished. Not one of these mass slaughterings was blamed on global warming, so why do journalists jump to report every modern endangered species status to be a result of global warming? I even read an article where the author attributed the El Niño Southern Oscillation to climate change. Too bad for the author that El Niño is recorded in the Peruvian Shelf marine sedimentary history for the past 20,000 years.

Whatever the reason for a predisposition to blame climate change, it is unlikely that every environmental concern is caused by climate change. We had these problems before climate change took over the news cycle, and they will still exist when climate hysteria is over. Here are my top 10 all-time picks for news stories that routinely fall short of excellence in reporting and fact-checking (11 if you include the claim that global warming is killing the Orcas). I have ranked them in order of the influence I think they have made in helping create the global climate hysteria we find ourselves in, with #1 having the greatest impact.

10) <u>Headline: CO2 is Pollution that Harms Human Health</u>

This is false. This news article often appears beneath a photograph of a crowded metropolis with the population wearing surgical masks to strain out the thick pollution. Recall that CO2 is a colourless gas and doesn't photograph well. The pollution in the photo is smog, and the people are wearing masks in an attempt to filter out the particulate matter suspended in the air that makes breathing difficult. CO2 is not a health hazard, but its anthropogenic release (in burning coal and diesel) is coincident with other noxious substances that are harmful. It's like drinking too much alcohol and saying it was the ice cube that gave you a headache. For every carbon dioxide molecule of human origin in the atmosphere, there are 32 molecules of carbon dioxide of natural origin. Increasing the CO2 content of the air that we breathe by 3% does not harm human health, as the building you work in probably has two or three times the CO2 than the outside air.

At the height of the 2020 Covid-19 pandemic, the worldwide industrial and transportation demands for electricity and petroleum products dropped significantly, by 20% and 25% respectively, and many global warming groups hailed the cleaner skies above China and India as evidence of what can be achieved by reducing carbon dioxide emissions. I trust that eventually science will prevail and it will be established that the cleaner air was due primarily to a reduction of diesel fuel used for transportation, which led to lower emissions of nitrogen oxides that cause the photochemical Los Angeles-type smog. The cleaner air was also due to a reduction of coal for industrial electricity generation, which reduced sulphur dioxide that causes sulphurous London-type smog. Both diesel and coal emit P10 and smaller particulate which also cause smog. I am confident that it will be determined that during the cleaner sky era of the 2020 Covid-19 pandemic, the atmospheric carbon dioxide levels remained unchanged or continued to increase. Air pollution dropped significantly, but not CO2.

9) <u>Headline: A Carbon Tax is the Only Way to Reduce Carbon Dioxide Emissions</u>

There are two issues with this. The first is that gasoline, or petrol, if you prefer, has a demand curve that economists call inelastic. That means the average consumer will still buy about the same amount of gasoline regardless of price, within a reasonable range. Consumers will do this because they need to drive or heat their home and are not wasting fuel because it is so expensive already. If the price of gasoline goes up or down, their diving habits don't change. The proof of this is in the observation that in the last decrease, the price of crude oil varied from $150 per barrel to $30 per barrel, and the demand for gasoline was fairly flat throughout. An exception to this would be when a car is replaced with a more fuel-efficient car due to high fuel prices. This is a slow process as the average car on the road in the USA today is 11 years old.

Another proof of inelastic demand is that the price for natural gas in home heating is at historic lows in North America, but homeowners are not turning up their thermostats from 20°C to 25°C because the heat is cheaper. They continue to heat their homes at the same comfortable temperature.

The second issue with the carbon tax solution is that in the past when the world had collectively made a huge mistake, like putting lead in gasoline that was slowly poisoning us, we simply banned the problem. We did not tax lead in gasoline, just as we did not tax chlorofluorocarbons when we discovered they were making a hole in the ozone layer. Nor did we tax DDT when it started killing the Condors. We banned lead, chlorofluorocarbons and DDT because we wanted to stop using them. We cannot ban the internal combustion engine, but if anthropogenic CO_2 is such a threat to our lives, our governments should mandate fuel efficiencies more stringently for larger passenger-carrying vehicles like SUVs and minivans. After the Arab oil embargo in the late 1970s, the USA used regulations to improve the average passenger car mileage from 13 miles per gallon to 28 miles per gallon in about a decade. Consumers responded by increasing the market share of light trucks from about 10% to today's 50%. The light truck category has lower fuel efficiency requirements and includes SUVs and minivans, both used primarily to carry passengers. If our governments genuinely want to reduce the amount of gasoline we use, all they need to do is apply the same stringent fuel economy standards to SUVs and light trucks.

A cynical person might suggest that our governments consider taxing CO_2 in gasoline *not* because they want less CO_2 in the air, but because they want more revenue, and they know we will accept the tax more easily if they sell it as "saving the planet." Yes, I just ripped off a plot from *Yes, Prime Minister*.

8) Video Opportunity: Glaciers Calving

Fortunately, this misleading video opportunity has seemed to have run its course and is no longer replayed as stock proof of global warming. The typical video was taken from a ship, normally near Greenland, and would show massive slabs of ice breaking off the edge of a glacier and crashing into the ocean. It is a process called calving, and it is unrelated to global warming. Glacial calving occurs because the glacier is growing, not shrinking, and the cause of growth may have occurred hundreds if not thousands of years ago. When snow falls on a glacier, typically more at the highest elevations, it compresses the previous snowfalls into ice, and the glacier gets thicker. The thicker part of the glacier moves by a slow plastic flow downhill toward the sea, hence the term glacial speed, and when it is pushed out over the edge of the land it loses the support underneath it. Sooner or later this cantilever of ice is too heavy, and it breaks off from the main glacier and crashes into the ocean. Someone floating by films this final growth spurt of a glacier and claims it is glacier shrinkage resulting from global warming.

Glaciers melt from higher temperatures, just like the ice on your sidewalk. Ice on the surface turns into water and runs downhill, which makes for a really boring video.

7) Photo Opportunity: Polar Bears are Starving Due to Global Warming

The sometimes-altered photo of the emaciated polar bear has also lost some of its popularity, even though at one point it was literally the poster image for the global warmers. One of the main reasons the public no longer blames a starving polar bear on global warming is the public realization that polar bears have no natural enemies, and unfortunately, at the end of a successful life, the typical polar bear death is to get sick or to become too old to hunt and then starve.

There are more reasons to dismiss the photo as proof polar bears are starving due to global warming. There is a polar bear population boom going on and how big this boom is depends on who you ask. The literature suggests that in 1960 there were about 10,000 polar bears worldwide, and they were threatened by overhunting. There is great debate on how many there are now, but 40,000 is not an unreasonable number, which is the midpoint of a range of estimates by Susan Crockford in her 2019 book

The Polar Bear Catastrophe That Never Happened. The World Wildlife Fund's 2018 population estimate was 26,000 polar bears. Again, it is not the intention of this book to referee between two opposing data sets, but rather to point out the logic underlying irrefutable trends. The debate should not be whether the polar bears have experienced a recent population increase in the range of 160% (World Wildlife Fund) or 300% (Crockford); the debate should be whether the polar bears are starving. I'm not a zoologist, but I don't think a quadrupling of any species population is a symptom of starvation. A reasonable explanation for the increased numbers of polar bears is that for the last 50 years, overhunting has been eliminated. This is a happier outcome than for the killer whale, bison or codfish, and underscores the polar bear's threat was not climate change but instead, the all too common result of encounters with man.

Polar bears have survived temperatures much warmer than what we have today. In the polar bear domain of northern Quebec, Canada, 130 km above the current tree line are the remains of a forest that grew from 3000 years ago and was destroyed by a fire about 400 years ago. The forest began in the Holocene Maximum and flourished throughout the Medieval Warm Period when the Arctic was about 5°C warmer than it is now. [53] However, it was the Little Ice Age that brought about its end.

6) Headline: Electric Vehicles Will Take Over the Market

Newsflash: According to the International Energy Agency, sport utility vehicle (SUV) sales on a worldwide basis have risen from 17% of the market in 2010 to 37% of the market in 2018. In that period, 165 million SUVs were sold compared to 5 million Electric Vehicles (EV), and electric and hybrid vehicle sales are now falling. The reasons drivers commonly give for not buying

---

[53]   Plimner, I: Heaven and Earth, p 403.

an EV are charging time, charging infrastructure, driving range, battery problems, and price when subsidies are removed.

While a plug-in hybrid electric vehicle can use 25% less gasoline than a sedan of the same model with a standard gasoline engine, stepping up in size from a sedan to an SUV can use 25% to 50% more fuel. With this trend, gasoline fuel demand should go up, not down. Half of all EV sales are in China, where the purpose is not to reduce CO2 emissions, but to replace gasoline demand with coal demand, the result of which national CO2 emissions in China will increase with the increased use of EVs. Inner-city air quality should improve because what were tailpipe emissions on the streets are now smokestack emissions outside the cities.

5) Headline: Global Warming is Causing Catastrophic Sea Level Rise and Islands Are Disappearing

Determining a changing sea level is very difficult because the land next to the sea could also be changing. Historical tidal gauges do not measure the absolute sea level; they measure the sea level relative to the land level. Plate tectonics can cause land to rise or sink, and the rebound of land, which was compressed by glaciers, continues today. Terrestrial glacier meltwater adds water volume to the oceans to cause a sea level rise; it also adds weight to compress the seabed lower, causing a sea level drop. Pumping out subsurface groundwater or petroleum can cause surface subsidence. Modern large coastal cities tend to cause sinking just by being there. Ancient seafaring migrants, who established civilizations on atolls in the Pacific and Indian oceans, would not have known at the time that an atoll is the top of a sinking volcano. The volcanoes don't stop sinking once people set up cities on them, and the rate of sinking can vary over time. Even though it's part of a continent, the Netherlands has been sinking for a thousand years. If you are in the UK, it is more complicated as Scotland is rising while parts of England are sinking. My Scottish grandmother would have liked that.

Determining if the global sea level is changing is problematic because the levels of dry land and the ocean beds are not static. The only sea level data that appear to be undisputed are satellite data as they measure the absolute sea level from a fixed orbit. Those satellite data tell us that from 1995 to 2019, the global sea level rose at 3.3 millimetres per year (about 13 inches per century).

We don't know if this is an accelerated rate because all data prior to the satellite data are based on the sea level with reference to the land, and land moves. We don't know how much, if any, of this post-1995 sea level rise is attributable to a 0.8°C temperature increase that began with the start of the industrial revolution, or whether the sea level rise is a normal trend, or if it's part of a normal cycle of increased and decreased rates of sea level rise. We just don't know. This uncertainty brings up the argument that if the ability of CO_2 to act as a greenhouse gas is now much diminished due to the logarithmic effect, then the likelihood that CO_2 can contribute to sea level rise is much diminished also.

If the current 25-year rate of sea level rise is slightly more than one foot per century, and we don't know if that is a trend or not, it seems a little alarmist to prepare for a three-foot sea level rise in 80 years, more than triple the current rate. You must admire the Dutch approach to sea level change; they have adapted to reality, avoided panic, and ignored doomsayers. Perhaps they have more confidence in surviving a sea level rise because they are so tall.

4) Headline: Forest Fires Are a Result of CO2 and Global Warming

What is sad about this headline is that not only is it false, but it is often printed while the fire in question is still raging, and the affected populations need help, not condemnation as authors of their own misfortune.

Two reasons the headline is false is that as a global trend, more CO_2 will make trees healthier, and warmer globally means wetter globally. These combine to make forests less susceptible to fire.

Two significant reasons for forest fires are drought conditions and forest management. In the Pacific Ocean, an El Niño event shifts the occurrence of rainfall from land to ocean, resulting in a drought. A similar pressure oscillation event in the Indian Ocean causes rainfall that would typically fall on Australia to fall in the ocean. Both these events predate the industrial revolution by tens of thousands of years, as droughts in North America and Australia. Of note, high surface temperatures on land do not cause drought, it's the other way around; drought causes high land surface temperatures. Recall that land has only 20% to 25% of the heat capacity of water. When there is water in the soil, the energy from the sun is partially absorbed as heat of vaporization to change the state of that water from liquid to gas. This can take a significant amount of energy and slow down the heating of the soil. When the liquid water is all gone, all the energy from the sun is used to quickly heat up the soil which then heats the air. Droughts set the conditions for forest fires, and droughts are connected to events like El Niño. We don't yet know what causes atmospheric pressure oscillation events over the oceans like El Niño, but since they have been occurring for tens of thousands of years, it is unlikely they will stop because of a tax on carbon dioxide or by adding ethanol to gasoline.

Forest management is another factor that contributes to forest fires. It is more than just clearing away the dead underbrush or allowing smaller fires to burn the accumulating deadfall fuel. In British Columbia, Canada, the government sprays some forests to kill deciduous trees so that more commercial coniferous trees will grow. The firefighters call the deciduous trees "asbestos trees" because they are highly resistant to fire compared to the conifers and can stop or slow down wildfires. Unfortunately, another cause of fires is that many are set intentionally, whether as acts of arson or to clear jungles for palm oil plantations.

It is wrong for media to blame global warming for forest fires because on a global scale, it is the opposite of what should happen with healthier trees in a wetter climate. It is very wrong for the media to divert attention away from drought, forest management, or intentionally set fires, as that is where the prevention of future fires will be found.

Wildfires do not occur where there is heat and wet; wildfires occur where there is fuel and drought.

3) Headline: Global Warming is Causing Extreme Weather

In the recent past, the press has claimed that global warming is linked to:

- Increased hurricanes and cyclones–This is incorrect. There has been no increase in hurricanes or cyclones. An analysis by the National Oceanic and Atmospheric Administration (NOAA) shows that the frequency of Atlantic tropical storms has been relatively constant since 1880. [54] The IPCC says they have low confidence there is any recent trend in the occurrence of tropical cyclones or extra-tropical cyclones in the last four decades, but also acknowledges that recent studies suggest a decreasing trend. [55]

- Increased tornadoes in North America–No change! NOAA data show the number of tornadoes in the USA from 1954 to 2014 is about the same, and the occurrence of more violent storms has slightly diminished. [56]

- Increased worldwide flooding from rivers–No again. The IPCC reported [57] that in the world's 200 major rivers since 1950, there was no significant overall trend. The number of reduced runoffs outnumbered the increased

---

54 gfdl.noaa.gov/historical-atlantic-hurricane-and-tropical-storm-records
55 IPCC, 2018: SR15 Chapter 3 table 3.2, and p 203.
56 ncdc.noaa.gov/climate-information/extreme-events/us-tornado-climatology/trends.
57 IPCC, 2018: SR15 Chapter 3 table 3.2 and p 201.

runoffs, and in each case they were consistent with local precipitation changes.

- Increased worldwide drought–Still negative. The IPCC says they have low confidence that there is a trend since 1950 in drought or dryness on a global scale, but acknowledges there were localized trends. [58]

One of the reasons we have the perception that extreme weather events are more frequent is our constant connectivity to instant information. Being better and more frequently informed doesn't mean something has changed. When my older son went to another city about 200 km away for university, the cell phone was still a rarity. If I inquired on a monthly telephone call how the food was, his answer was, "Meh!" When my younger son, who had a cell phone, was in university in a different city 14,000 km away, I could and did find out by a photo sent by text what the food was like, and maybe several times a month. Both sons kept up their normal intake of food, but the younger one sent me more information from further away and more often because he could do so cheaply. It's the same with the weather. The statistics are about the same, but our awareness through information technology is much higher.

I have often thought that our increased urbanization has led to a lesser connection to nature and a false belief that the weather is normally benign. When you commute to your job in an office building by car or train, you don't experience weather on the same level as someone growing up on a farm in Saskatchewan who rides a horse bareback–and barefoot–10 miles to school, as my dad did in the 1930s. ("Uphill both ways! In my pyjamas during a blizzard!") Maybe in our comfortable, insulated and latte-filled lives, we are more surprised than previous generations when we do

---

58   Ibid: 3 table 3.2, and p 196.

experience bad weather. The statistics do not support any claim that bad weather is more frequent.

2) Headline: Warmest Decade/Year/Season - Ever

This is probably the one type of media story that polarizes the climate change debate the most. Headlines scream that the hottest event ever just happened, then politicians and climate activists repeat the story and declare climate emergencies in the belief that if repeated enough, it will become true. Normally they are not outright lies or fabrications, but they are hardly ever truthful. There are several ways the media and politicians can deliberately mislead you and get away with it, and here are the questions you can ask to protect yourself from misinformation.

i. Who is the originator of the temperature database?

There are many originators, and all of them may have a bias toward a political viewpoint on anthropogenic climate change. The databases that are used exclusively by pro-IPCC or anti-IPCC groups may be more likely to have a bias. Databases that are common to both groups may have more independence. A good example of bias being exercised is the treatment of the urban heat island effect. Large cities generate and store heat by their buildings, asphalt, internal combustion engines and industry. If a small town in 1900 becomes a large city by 2000, how is the temperature reading in the year 2000 adjusted to make it comparable to the temperature reading in the year 1900? A city of 1 million could have a daytime heat island effect from 1°C to 3°C and a night-time effect of up to 12°C. Take the time to look up two or three temperature databases and see if they agree. I'll offer some comments on the three most recognized global temperature data sets.

The most well-known is HadCRUT4. This is the database we discussed extensively in Chapter 4, and which is a cornerstone for the IPCC. While it survived the Climategate accusations of wrongdoing, it has never overcome its reputation for bias toward the IPCC agenda, and its claims of providing a global average temperature back to 1900 are greatly overstated.

Another widely quoted global temperature database is NASA/NOAA. As mentioned earlier, NASA's website is very supportive of the IPCC viewpoint but does not independently verify IPCC's work. The NOAA is where the 2015 whistleblower incident originated, with claims of exaggerating temperature increases just before the signing of the 2015 Paris Agreement. However, the NASA database now includes the land-based US Climate Reference Network, which is showing a slight recent cooling in the continental USA.

The third most known global temperature database is from the University of Alabama at Huntsville (UAH), which works with NASA/NOAA satellite data. They also incorporate weather balloon data with the satellite data to measure the temperature in the troposphere. They do this because that is where the IPCC models predict the first global warming "hotspots" will occur. Their prominence in the debate occurred when researchers John Christy and Roy Spencer [59] found that there are no atmospheric hotspots (another failure of the IPCC climate change models), and that errors in the orbits of some satellites caused an overstatement of previous NASA

[59]   Christy J., Spencer R. 2018: *Examination of space-based bulk atmospheric temperatures used in climate research.* International Journal of Remote Sensing, Volume 39 Issue 11.

global temperature surveys. UAH calculates a much smaller temperature increase in the lower atmosphere in the tropics (0.10°C per decade) than the IPCC (0.27°C per decade) for the period from 1979 to 2016.

ii. What is the time frame of the database being quoted?

Often the fine print in the article will state that the "hottest decade in history" really means the "hottest decade in the last five decades."

iii. What data trending statistical methods have been applied?

As Samuel Clemens (pseudonym Mark Twain) complained, there are "lies, damned lies, and statistics." All databases will take the raw data and apply statistical techniques to smooth the variations and search for trends. If that organization already prefers the trend it would like to see, they will choose a statistical methodology and exclusion of data points that support the trend they want (confirmation bias again).

iv. What is the precision of the temperature increase compared to the accuracy of the thermometer?

If someone claims that it is 0.5°C hotter today than it was 100 years ago, and the accuracy of the 100-year-old thermometer was then 1°C, it is possible that the temperatures today are anywhere from 0.5°C colder to 1.5°C warmer.

Here is a good example of how the confirmation bias and presentation of data can make a difference in the headline. In January of 2020, NASA declared that 2019 was the second warmest year on record by 0.95°C. Here are the misleading parts. The headlines don't explain that it is the warmest year since only 1880. So again, the fact that it has been warmer than today for 90% of the time over the last 12,000 years has been omitted. And as previously discussed, the headline glosses over the fact that we don't know what the global average temperature was in 1880.

The headlines also neglect to explain the degree of uncertainly there is in estimating the world's average temperature for 140 years to the nearest 1/100 of a degree. There is no discussion that if it wasn't for NASA informing us, it is unlikely anyone on Earth could have noticed a 1°C temperature increase in a normal lifetime, let alone 140 years. What they certainly don't highlight is that the most technologically advanced and comprehensive ground-based temperature recordings in the world, which are now part of the NASA database, show that the continental USA has cooled by about 0.4°C since 2005. The US Climate Reference Network was set up to address the issues of temperature trends in the USA, and it's apparently not warming up. It seems odd that the only part of the world that is not warming is where the most advanced temperature recordings are being taken, and in the rest of the world where there are poorer data, it is warming. [60]

My favourite example of all this is the hottest temperature ever recorded in Canada: 113°F (45°C) on July 5, 1937, at Yellow Grass, Saskatchewan. My dad was an almost nine-year-old boy on a farm 10 miles away and remembers the exact day because he was disappointed that the community baseball game was cancelled due to the heat. Yet modern databases that promote global warming constantly tell my dad it is hotter in Canada now than it was when he was a kid. He is 92 and cantankerous about events in his life being rewritten, especially since he grew up in the

---

60   wattsupwiththat.com/2020/01/15/while-noaa-nasa-claims-2019-as-the-second-warmest-year-ever-other-data-shows-2019-cooler-than-2005-for-usa/

epicentre of the Dirty Thirties (which he occasionally mentions were hot and hungry). I explained to him that the data came from a university in the UK that also records a town in Columbia with temperatures of 80°C for three months, and that he is older than the records they are looking at as a baseline. Furthermore, they are pro-global-warming-caused-by-man, so they manipulate the data and thermometer accuracy to get rid of the spikes and to get the temperature trends they want. They then back extrapolate that trend to cover the period from 1900 to 1950. That is why 113°F on July 5, 1937, the hottest day in Canada ever recorded, is no longer considered a relevant scientific fact as it has been superseded for a convenient political campaign. Fortunately, the town of Yellow Grass, Saskatchewan is still 999,629 people short of 1 million, so I don't have to explain the urban heat island effect also. It is no surprise my dad is a fan of Mark Twain.

Figure #13: Walter Barmby and Family, circa 1934, Lang, Saskatchewan

July 5, 1937 was the hottest day ever recorded in Canada: 113°F (45°C). My dad was there to witness it three days before his ninth birthday. He's the handsome boy in this picture, which was taken on the same farm three years earlier. (61)

My dad told me the story about that summer of 1937; it was the only year during the Great Depression that not a single drop of rain fell on his dad's wheat farm in Lang, Saskatchewan. The reason it was the hottest day ever in Canada was that there was no water left in the soil, and all the sun's energy was available to heat the soil and radiate upward. There were no forest fires in Lang that year, because as the saying goes in southern Saskatchewan, "Behind every tree, there is a beautiful woman."

1) <u>Political Campaign Speech: 97% Scientific Consensus on IPCC Conclusions</u>

This is undoubtedly the most successful misinformation campaign by the global warmers, and it also displays either the pro-global warming bias of the press or their indifference to fact-checking. I hope I have made the case that the science is *not* settled, especially since the IPCC scientists readily admit that they have not figured it all out, and their predictions are consistently

---

61 Here's a little family history. Walter Barmby not only outlasted the Dirty Thirties, he and a small group of other farmers started Federated Co-operatives Ltd., now a very large company. To break the big oil stranglehold on farm fuel prices, they started the only co-operatively owned oil refinery in the world, still operating today in Regina. I often lament that if Granddad was a capitalist instead of a socialist, I may have been born with a silver spoon. His wife, Marjorie (nee Wight) was an 11th generation American, and my dad married an 11th generation Canadian. That makes my grandchildren 14th generation Canadians on my mom's side (nee Ball), and 15th generation North Americans on my dad's side; five different centuries in North America on two branches of the family tree.

overstated. The Summary Reports for Policy Makers written with veto power by the UN political appointees are intended to minimize any the doubt that the contributing scientists may have, to minimize the discussion of any limiting factors to the greenhouse gas effect, and to simply change the discussion to a different focus rather than address uncontestable deceptions like the hockey stick graph. The IPCC Summary Reports appear to be well-documented validation of scientific consensus.

Let's briefly investigate how an overwhelming consensus can be manufactured. At least three major brands of toothpaste have been advertised, sometimes simultaneously, as "recommended by nine out of 10 dentists." It is, of course, impossible for 90% of all dentists in any one country to recommend one specific toothpaste brand, and then apply that claim on three different brands. All three brands got away with it by conducting their own research and then easily manipulating the data. Ask a question like, "Do you recommend to your patients to brush with toothpaste?" and then add a second question, "If you recommend brushing with toothpaste, would Crest/Colgate/Sensodyne be okay?" If nine out of 10 dentists said "yes" to both questions, then all three toothpastes would have survey results saying their brand is implicitly recommended by nine out of 10 dentists. Eventually, the impact of the 90% claim became ineffective, and the brands switched their advertising.

A second item to consider is that the media has never questioned the improbability of the near-unanimous concurrence of what must be a very large and diverse group: all the scientists in the world who have an opinion on climate change. Even if you manipulated the questions in an attempt to claim consensus on the answer, a 97% agreement rate among free, independent and well-educated people is unlikely. 97% is more like an election result in Turkmenistan, and I know because I have worked there.

I made many trips to Turkmenistan, and I was one of the few foreigners who were granted permission to leave the capital of Ashgabat. On a trip to the Caspian port city of Turkmenbashi to inspect ships suitable for charter, I think I witnessed the only Stalinist state I may ever see in my lifetime. Turkmenbashi was very different from the showcase capital city. There were no golden statues, no marble facades on apartment buildings, and no impeccably clean wide paved boulevards. With my Turkmen companions and the members of the government agencies I met, I was free to discuss religion in a 100% Muslim country. We agreed that my God and their God was the same as Abraham's God, so all was good. We could discuss sports and music, and amazingly, the Canadian rock band Nickelback was a huge hit on the black market. I promised to send back CDs that I now realize I never did. I was not allowed to discuss politics, and there was instant paranoia even at the hint of that subject. Rooms went silent, ears perked up, and guards started coughing a lot. Politics and fear were the same thing.

97% was the actual election result in Turkmenistan in 2017, where the opposition candidates were appointed by the government. The western media called this election a sham. Strangely, the western media not only accepted but also promoted a 97% positive endorsement of the IPCC climate change thesis in what must have been a massive survey for which no record has ever been produced, and no idea of which scientists were deemed sufficiently qualified to take the phantom survey.

Had the public or the media asked those questions, they would have discovered that no such survey ever existed. Like World War II propaganda, the practice of the global warmers has been to repeat the 97% scientific consensus message so many times that the public has deemed it to be true.

You may have noticed that up to now, I have mentioned names of individuals only from a historical, pre-global warming context, or names in the footnotes where fellow authors have contributed their views to the public domain. I did this thoughtfully as my intent is to argue the issues of global warming, not attack the personalities taking sides on them. Too much of the global warming debate has devolved into personal attacks rather than rational discussions of differences. I will now make an exception to that philosophy as these two individuals and their statements are so famous that they cannot be hidden: Al Gore and Barack Obama.

In 2004 a study was conducted [62] that reviewed the abstracts of 928 scientific papers published between 1993 and 2003 based on the search words "climate change." The study found that 75% of these papers " either explicitly or implicitly accepting the consensus view claimed by the IPCC that human greenhouse gas additions were causing most of the temperature increases over the last 50 years," and that none of the papers directly disagreed. Here are a few constraints on extrapolating these results further. This is not a survey of all qualified persons who have an opinion on climate change; it is only a reflection of those who had published under the subject of climate change, and for example, could exclude papers on solar variation and the effect of cloud formation on Earth. Reading the abstract of a paper is not the same as reading the full paper; the abstract is only a summary of the main conclusions and omits a lot of detail. The term "implicitly accepting" is subjective, even more so if the study author only read the abstract.

In his 2006 movie, *An Inconvenient Truth,* Al Gore made a reference to "...a massive study of every scientific article in a peer-reviewed journal written on global warming for the last 10 years..." He continued by saying "...and they took a big sample of 10%, 928 articles..." In the same sentence, Gore asked, "...you know the number of those who disagreed with the scientific consensus that we're causing global warming and that it's a serious problem? Out of 928, zero."

It is a reasonable assumption that Gore's reference to the 928 articles with "zero disagreements" links to a statement in

---

62   Oreskes, N. et al *Science* 03 Dec 2004: Vol. 306, Issue 5702, pp. 1686

the aforementioned 2004 study quoting 928 articles with "none directly disagreeing." If this is correct, then Gore exaggerated when he stated that the 928 articles were only 10% of a much larger database, which implies there were 9280 papers. He then led his audience to believe that in this much larger study, there was no disagreement that humans are the cause of global warming and that it is a serious problem. In reality, only 696 papers (75% of 928 papers) within a narrow search category of only two words, and based on only the abstracts of the papers, agreed–explicitly or implicitly–that human greenhouse gases had caused half of the temperature increase in the last 50 years.

An Inconvenient Truth launched three major untruths into the international consciousness: the exaggeration that there was near-universal scientific consensus with the IPCC view that climate change was mostly man-made from carbon dioxide emissions; the fear that climate change is catastrophic; and the deception of the hockey stick diagram that negated 1000 years of naturally occurring climate change and eliminated the warmer medieval temperatures.

Gore also predicted in the movie that the glaciers on Mt. Kilimanjaro would be gone within a decade, so in 2016 my wife and I climbed up to check. The glaciers were still there. Our guide told us that 80% of the glaciers had already disappeared in the time between the publishing of Ernest Hemingway's short story, *The Snows of Kilimanjaro,* and Al Gore's prediction. Gore was betting on a horse already way in the lead, and he still lost. The shrinkage of the glaciers was caused by deforestation all around Mt. Kilimanjaro, which reduced the humidity on the mountain and slowly caused the ice to shrink by sublimation. Sublimation is the same effect as ice cubes shrinking in your freezer; it's still cold, but the ice transfers directly into a vapour without melting due to the low humidity.

Obama made use of similar political rhetoric, and with significant embellishment. On May 15, 2013, a study [63] was released of 11,944 scientific papers that met the search words "global climate change" or "global warming." The study found that 4014 papers had expressed an opinion on whether global warming was caused by the human release of greenhouse gases. Again, using only the *abstracts* of these 4014 papers which expressed an opinion, 3896 papers had either explicit or implicit endorsements that humans are causing global warming. These 3896 explicit or implicit endorsements of the IPCC climate change view represent 97% of the 4014 papers in the study that expressed an opinion, but only 34% of the *total* study of 11,944 papers. The abstract of the paper (which was a study of abstracts of other papers) stated, "Among abstracts expressing a position on AGW, 97.1% endorsed the consensus position that humans are causing global warming."

The next day, President Obama tweeted to his 31 million followers that "Ninety-seven percent of scientists agree: #climate change is real, man-made, and dangerous." Obama did not say where his data came from, but as with Gore, is seems reasonable that he quoted the 97% headline from one day later.

If Obama was referencing the study released the day before, he should have noted that the study never said climate change was dangerous. He would have also known that 3896 papers do not constitute 97% of scientists in the world, as his tweet implied. He should have mentioned that 66.4% of the papers reserved any opinion on the IPCC viewpoint. Assuming Obama or his staff read the paper before he tweeted, he should have also noticed that the 2013 study included the same 928 papers referenced in the 2004 study. The 2004 study had concluded a 75% consensus of the IPCC view, which Gore embellished into a unanimous consensus. The May 15, 2013 study reviewed the same 928 abstracts for explicit and implicit agreement and determined

---

[63] Cook, J. 2013: Quantifying the consensus on anthropogenic global warming in the scientific literature, Environmental Research.\ Letters. 8 (2013).

that only 894 of the papers qualified for the study. Of those, only 28% supported the IPCC viewpoint, and 72% expressed no opinion. Obama should have been concerned that the reference in *An Inconvenient Truth* to an implied 9280 papers on climate change where none disagreed with the statement "We're causing global warming and that it's a serious problem..." was in fact only 250 papers (about 3% of Gore's implied number).

Obama should have been concerned because the 2013 study greatly undermined the 2004 study quoted by Gore, which highlights the potential of confirmation bias in determining what constitutes implicit support. This presents an invitation to eliminate subjective implicit support and rely only on objective explicit support, which is not good from a politician's perspective. This did happen, as a few months later a review of the May 15, 2013 paper was released. [64] Dr. David Legates accessed the public domain data of the May 15 paper and discovered that only 41 of the 11,944 papers explicitly agreed that mankind was responsible for at least 50% of the global temperature increase since 1950 (the IPCC view). Of the 97% "consensus" of explicit or implicit support of the IPCC view claimed by the authors of the May 15, 2013 study:

- 0.3% had stated they explicitly agreed with the IPCC,
- 1.0% expressed an opinion that they disagreed with the IPCC,
- 32.6% were subjectively deemed to implicitly support the IPCC view, and
- 66.4% expressed no opinion.

Ironically, these were the same authors who revised the 75% consensus of the 2004 paper down to 28%. As far as climate change being dangerous, that appears to be Obama's add-on.

---

[64] Legate, David. 2013. Climate Consensus and 'Misinformation': A Rejoinder to 'Agnotology, Scientific Consensus, and the Teaching and Learning of Climate Change'

The former Vice President of the USA gave credibility to the claim of near-unanimous scientific consensus of the IPCC position in a survey. (You will recall that he implied the survey was based on more than 9000 papers, and that the May 15, 2013 study revealed that only 250 of the 894 qualifying papers were supportive of the IPCC position.) Years later, the sitting President of the USA cemented that credibility with a claim of 97% scientific consensus that "...#climate change is real..." the day after the May 15, 2013 study of almost 12,000 papers was released. The findings of that study were very quickly challenged as only being 0.3%–41 papers–expressing explicit support for the IPCC.

Compare this to a more direct, objective, non-interpretive approach—the Oregon Petition. It was a clear, unequivocal statement urging the USA not to sign the 1997 Kyoto Agreement that was signed by 31,487 American scientists of whom 9,029 held PhDs. Every signature on that petition is available to the public. The next time someone says that 97% of the scientists in the world agree with the IPCC, ask them to produce the list of scientists. Otherwise, you are falling for the oldest and most effective propaganda trick known to man: repetition of a lie leads to its acceptance as a truth.

After both Gore and Obama twisted the truth, threw in fear, and offered their leadership as hope, a concerned world accepted as fact that man-made global warming was real and dangerous, and placed their faith in a scientific consensus that did not exist. Only 41 papers explicitly stated that mankind is responsible for the release of greenhouse gases, and that those gases have caused at least half of the increase in the world's temperature since 1950. Furthermore, not one of the papers said greenhouse gases were dangerous. Almost three times as many papers, 118 in total, went on record disagreeing with the IPCC. Neither group can be considered a consensus.

CHAPTER 7

Saving the Planet

The Paris Agreement Will Fail

The objective of the 2015 Paris Agreement is to reduce the emissions of greenhouse gases. It is doomed to fail because of three major flaws.

The first flaw is that the biggest greenhouse gas emitters in the world have either withdrawn, reconsidered, or are now perversely incentivized to maximize their CO_2 emissions in the next 10 years. China contributes 23.75% of global greenhouse gases and has pledged to peak the emissions by 2030. India, who accounts for 5.73% of the greenhouse gas emissions, has not pledged an absolute reduction, but rather to reduce them only per unit of GDP. This means that 29.48% of the worldwide human greenhouse gas sources have a green light to get bigger until 2030. The 2017 to 2018 greenhouse gas emission increase of these two countries combined is larger than the total greenhouse gas emission output of the United Kingdom in 2018. [65] If they are eventually compelled to reduce their GHG emission in 2031, it will be easier to do so if they can maximize their CO_2 emissions before 2030.

---

[65]   Climateactiontracker.org/countries

The next largest greenhouse gas emitter is the USA at 12.1% of the world's total. They have withdrawn from the Paris Agreement.

The European Union contributes 8.97% of the global total of greenhouse gas emissions, but the biggest greenhouse gas emitter has new plans. Germany has decided not to close its remaining coal-fired electric power generating plants by 2030, but to keep them running to 2038. As these coal-fired plants contribute 7% of the EU's total emissions, it is highly unlikely the EU will meet their pledge to cut 40% of its emissions by 2030.

Let's take a quick look at the next six big emitters to round out the top 10. Brazil (5.70% of emissions) pledges a 37% reduction; Russia (5.35% of emissions) signed up for a 25% to 30% reduction; Japan (2.82% of emissions) has a 26% reduction commitment; Canada (1.96% of emissions) pledges a 30% reduction; The Democratic Republic of the Congo (1.53% of emissions) and Indonesia (1.49% of emissions) both say they will try harder. These six countries represent 18.85% of the world's greenhouse gas emissions, and four of them made hard reduction commitments. None of these four is on track to meet their pledges.

According to Climate Action Tracker, the first country on the list with a firm reduction in emissions that is likely to be met and that is compatible with the 2°C warming level of the Paris Agreement is Ethiopia, who makes up 0.35% of the worlds emissions and pledged a 3% cut. The first and only country on the list with a firm reduction in emissions that is likely to be met and is compatible with the 1.5°C warming level of Paris Agreement is The Gambia, who makes up 0.01% of the world's greenhouse gas emissions and pledged a 45% cut to some of its economic sectors.

I worked in several African countries, including The Gambia. It is the smallest nation on continental Africa, an enclave of Senegal and sadly, once a major centre of the slave trade. The Gambia struck me as an African symbol of hope. It is a fiercely independent and poor country, more democratic than most in the developing world, with proud and friendly people. As I was there on one visit

with a former Prime Minister of Canada, Mr. Jean Chretien, I got to meet the President. We visited the village of Kunta Kinte of *Roots* fame, and the island prison where the slaves were held before transport. I made a friend in the government there who had heard of my business involvement in running a hand-wash carwash to give jobs to unskilled street kids, and he wanted to do the same in the capital, Banjul. He never got the carwash going because he died in hospital of a treatable kidney infection. He wasn't treated because the hospital had no electricity at that time.

The Gambia does not receive any benefit by being the only one of 195 original signatories to be currently reducing emissions as they committed to do under the most stringent 1.5°C warming Paris Agreement targets. The Gambia needs exactly the opposite of what the Paris Agreement offers; it needs cheap electricity so that its citizens don't die in hospital waiting for the power to be restored. Alhamdulillah, Mr. Siaka Camara, I believe your country misses you.

Failure number one of the Paris Agreement is that the three largest emitting signatories plus the European Union, representing 50.55% of global greenhouse gas emissions, are either on a path to increase emissions substantially, won't meet their emission reduction targets, or no longer care that they had an emission reduction target to begin with. The next six largest greenhouse gas emitters bring the aggregate of all nine countries, plus the European Union, to 69.4% of the world's total, and not one of them has a meaningful reduction target that is on track to be met. As of this writing, 99.99% of the world's greenhouse gas emissions are from countries not on target to meet the reductions needed for the 1.5°C warming target of the Paris Agreement. Let me suggest that this 99.99% failure rate does not indicate indifference to pollution or climate change, but rather a measure of skepticism of the 194 remaining signatories that climate change is completely man-made and dangerous. (Might one call this heresy?) Or it could be, as my good friend Carl Hodge, a professor of political

science at the University of British Columbia, explained to me: as a general rule, the sincerity that governments bring to any multilateral agreement is inversely proportional to the number of governments that sign it.

The second flaw that is going to cause the Paris Agreement to fail is the promise to ramp up existing Green Climate Fund transfers from the wealthy countries to the developing countries by 2020 to $100 billion per year and continue that level through to the end of 2025. The USA, representing almost 25% of the world's GDP, has left the club and is irreplaceable for meeting this goal. Other rich countries may find it difficult to persuade their taxpayers to both develop both green energy and contribute to this fund, and it will be increasingly less affordable as the cost of a unit of GDP will go up by the cost of green energy. The Yellow Vest protests in Paris may be a bad omen for the 2015 Paris Agreement. Taxpayers are being asked to increase their cost of living to reduce greenhouse gas emissions and send more money to developing countries to help reduce *their* greenhouse gas emissions, while on a global scale, the big emitters emit more. Something is going to give.

Failure number two in the Paris Agreement is that those rich economies that make every effort to meet their emission reduction targets are going to be treated like lambs. First, their economies will be fleeced and then slaughtered. Developing countries will further prosper with coal produced electricity. Developing countries will receive subsidies from the rich, which may be justified as long as the rich are burning food for fuel for no apparent environmental gain.

The third flaw is that the commitment to the Paris Agreement from countries with greenhouse gas emission reduction targets will be further questioned when it is revealed that the Agreement in fact incentivizes the burning of coal to generate electricity. Put yourself in China's or India's shoes. Global warming has been insignificant since 1998, and advanced temperature recordings

established in the USA have shown slight cooling in that country since 2005. If the atmospheric content of CO_2 reaches 560 parts per million by 2030, and the global temperature measure by an unbiased independent database has not risen much from 1998 (aided by the weakest 11-year solar cycle in two centuries predicted by NASA), then they can rightfully claim two reasons to carry on emitting. Firstly, the IPCC models were too sensitive to CO_2, and secondly, we may have hit peak global warming because of the logarithmic relationship between CO_2 and temperature (which the IPCC concedes). Add to that the Paris Agreement rules that fail to promote exporting liquefied natural gas that is lower in CO_2, less sulphurous, and lower in particulate matter to other countries that burn coal. This further promotes the burning of coal.

In 2030, China and India may also argue that there is a stronger correlation between solar activity than carbon dioxide to global temperature, effectively arguing that the theory proposed by Svante Arrhenius in 1896 has been disproven. Other issues will weaken the commitment of countries to the Paris Agreement. For example, those countries who have previously invested heavily in hydropower will object to changing the rules mid-game when the hydroelectric reservoirs, which emit no CO_2, will now be classified as significant emitters of methane.

Failure number three in the Paris Agreement is that it is a lopsided deal that favours economies currently growing on coal. If the IPCC forecasts upon which the Agreement is based are not realized, those economies will stick with coal permanently. The worst-case scenario for the Paris Agreement will be people looking out their frost-covered windows and joking, "Remember Y2K? Remember global warming?"

A Plan for Saving the Planet

I hope it will not take until 2030 for the people of the world to realize there is no climate emergency. In the next 10 years, the United Nations may try to save the Paris Agreement, and governments may try to save their economies. Nevertheless, we will still need a plan to save the planet, because the real tragedy is the global environmental degradation that is continuing while our attention is laser-focused on global warming. I have some suggestions, but first, I need to look in the mirror.

Reality Check:

You are now an old boomer who, after a modestly successful engineering career that spanned the globe, is enjoying a comfortable retirement by a beautiful lake in one of the most desirable countries that ever existed. You know that the success of this writing will be measured by the amount of grief that comes raining down on you and your family from the billionaire-funded climate change industry and its political faithful. It is not them you want to reach. Your goal is to reach reasonable people and ask them to think reasonably about climate change. Remind yourself that you are writing for your grandchildren's futures. You don't want them to be intimidated by the fear of something you know is mostly hysteria, and you want them to have the choices in life that you did. Your grandfather did not give up in 1937; instead, he doubled his efforts on the farm.

Why are you so sure that climate change is part natural and part man-made, that the man-made portion is over or nearly over, and that there is at least an equal chance the planet is beginning a naturally cooling period?

- The first IPCC report acknowledged that the Earth's temperature changed naturally, and the current temperatures are significantly lower than most of the recent human era.
- The last IPCC report acknowledges that the rate of temperature increase slowed from 1998 to 2012 (it

stopped, and the best land-sourced data available show the continental USA cooling a bit since 2005).

- The atmospheric hot spots the IPCC predicted would appear in the tropic regions continue to be absent while human CO_2 emissions are growing significantly.
- You have uncovered in the IPCC files that CO_2 has a limited ability to act as a greenhouse gas, and the temperature trends acknowledged by the IPCC and satellite data suggest we are near that limit. This is a Catch-22 for the global warming community, who insist we must accept the science from the IPCC.
- A Grand Solar Maximum dominated the 20th century. NASA predicts the current solar cycle will be the weakest level of solar activity in 200 years, and prominent geophysicists are predicting the 21st century will be dominated by a Grand Solar Minimum. There is a very strong theoretical and observed relationship between solar activity and climate change, and the IPCC has dismissed this.
- As I write this from a log cabin in the interior of British Columbia, Canada (some refer to it as British California because of its leftist politics) during a brutal winter cold snap, my weather app just flashed that 100-year-old record cold temperatures are being broken.
- I am not alone. Five years into a 15-year window, only one minor signatory of the 2015 Paris Agreement with CO_2 reduction commitments is meeting the 1.5°C target. Perhaps the other signatories followed the same logic I have described and came to similar conclusions, or perhaps global warming is not their biggest problem. It seems to me this fatal failure rate signals a different agreement among them: there is no climate emergency.

I am confident in my position because I have been trained to stand on the shoulders of giants, and those giants are telling me, "It's the sun, stupid." Here are my humble suggestions for saving the planet.

To the United Nations:

Disband the IPCC; it is hopelessly compromised and irredeemable. Here is a short recap of two major failings, starting with the hockey stick graph debacle:

- The computer model that produced the 1999 hockey stick graph was investigated by the US government and found to inappropriately produce hockey stick shapes.
- The bristlecone pine tree ring data used in the study was known to be sensitive to CO_2 fertilization.
- The 2001 IPCC report suppressed data that would have raised concerns about the bristlecone pine data.
- This was not revealed until the 2009 Climategate email hacking release, after which the hockey stick graph disappeared (although the hockey stick mentality remains).

And then the opinions on future temperature increases:

- The climate change computer forecasting models have consistently overstated global warming, indicating over-sensitivity to carbon dioxide.
- The IPCC forecasts completely missed a cessation of global warming from 1998 to 2012, indicating there are some non-CO_2 factors at play which they disregarded.

Accept that the work is biased by the mandate you wrote. Transfer all the work done to the United Nations Educational, Scientific, and Cultural Organization to salvage what is worth salvaging. Let's not throw the baby out with the bathwater. I am suggesting this as UNESCO took the lead in identifying that hydroelectricity produces significant amounts of methane, something the IPCC should have done themselves. Build the best global temperature database possible and audit it regularly. Conduct an audit of the most recent IPCC work with a view to introducing diversity of scientific disciplines and views, ensuring the facts are used in an unbiased manner, and disclosing the limits of our understanding. This would include investigating the relationship between the sun's cycles and the Earth's climate cycles and constructing global temperature estimates that don't change historical fact. And, since computer modelling of forecasting climate has had such poor results, why not try the artificial intelligence technique of mining the past to see if there are predictable patterns. Be upfront that the greenhouse gas effect is not sensitive to each increase in atmospheric CO_2, but to each doubling of atmospheric CO_2. Make this work available as a library.

To the Media:

Stop dismissing the experiences of real people. My dad is part of the greatest generation and deserves some respect. He lived through hotter times on the Great Plains in the 1930s and worked releasing weather balloons near the Arctic Circle in the warmer early 1950s. Don't dismiss these experiences; explain them in context.

Don't mislead and scare another generation, like some of today's teens who have now lost hope. That generation is gifted with the longest life expectancy in the most peaceful and prosperous world ever, and yet some don't even want to finish high school because they fear climate change. Your alarmist reporting with childlike cartoons of dumbed-down science helped create the current great depression, which is a generation fixated on extinction. Saturation coverage to celebrity activists should be at least balanced with coverage of reputable scientists who have put their careers on the line as whistleblowers. You should be encouraging debate and seeking truth, which is your traditional role in the Age of Enlightenment and is essential to our democracies. Help us by example to restore civil free speech.

To our Politicians:

When global warming either stops or becomes a slow and negligible change, there will still be global environmental challenges to face, such as overfishing, plastics in the oceans, habitat destruction, and trans-border pollution in all its forms. We elect you to represent us to solve these issues. You are not elected by radical and vocal special interest groups to invent a global and unaccountable jurisdiction to dump these problems into with a mandate that predetermines the outcome. Stop treating the IPCC as if it were the oracle of climate science; they are unaudited, politically motivated, tainted by scandal and whistleblowers, and their unscientific methods have a long track record of hyperbolic exaggerations and grossly missed forecasts. Will they still be unassailable when biogenic methane from hydroelectric reservoirs gets added to your national greenhouse gas emissions?

In our own jurisdictions, we need to clean the air above our cities of particulate matter, sulphurous smog and photochemical smog. If we are causing flora or fauna extinctions, we should address those issues scientifically, not blame all environmental calamities on a single trace molecule that all life on Earth needs. Be careful how far and how fast you push the political pendulum just to get re-elected. If it slips from your hands, it will swing just as far back in the reverse direction, and no net progress will be made. Declaring a climate change emergency is fearmongering for political gain at the expense of our children. You are more culpable than the media as you encouraged fear to be part of the educational curriculum. Make sure our schools teach science (and the Scientific Method), not the viewpoints of political activists.

The judge in Chapter 1 never did recall the jury for their verdict. As your constituents were part of the jury, I would invite them to consider the verdict on the next national election ballot. They may now take the heretical view that carbon dioxide is not causing environmental destruction. You cannot fight naturally caused climate change, but you can fight pollution.

To the Reader:

I hope this book has helped you prepare for the cult-like membership question "Don't you believe in climate change?" You could deflect the question by asking if they can explain the Second Law of Thermodynamics and how it always warms the Arctic more and makes teenagers' bedrooms messy. That's generally a conversation stopper. Let's say they are not easily diverted and have teenagers and know about entropy, yet still want to know why you are not a member of the club that everyone else seems to have joined and are fanatics about.

When someone tells you the global temperature is increasing, ask them where their data are from, and if they can explain the pause in global warming from 1998 to 2012.

If they claim there is a 97% consensus among scientists that global warming is man-made and a serious problem, ask if they have seen the latest and exciting election results from Turkmenistan.

When they say the United Nations Intergovernmental Panel on Climate Change has scientifically proven that we must limit CO_2 emissions to keep the global temperature increase to 1.5°C by 2030, ask if they know:

- that water vapour creates 90% of the greenhouse gas effect;
- that for 90% of the time in the last 10,000 years, it has been warmer than today, while in the last 500 million years there has been no relationship between CO_2 and temperature;

- that the IPCC admits that the concentration of CO2 in the atmosphere is not the key factor, but it is the doubling of the CO2 concentration that is important.
- the IPCC's main temperature data set is full of obvious errors, and wasn't global before 1950;
- whether they can produce an IPCC forecast that has proven to be correct;
- how long the IPCC whistleblower lineup is; and
- what happened to the IPCC hockey stick graph.

If they insist that the 2015 Paris Agreement to reduce greenhouse gas emissions has the full support of all 195 signatories, ask them:

- why only a handful of those countries have meaningful greenhouse gas reduction targets;
- why China and India combined can increase their greenhouse gas emissions in one year by the same amount the United Kingdom produces in a full year, and still be in compliance with the agreement;
- why there is a 99.99% failure rate to meet the 1.5°C warming target;
- why they support it if it promotes coal usage, which contributes to the particulate matter in the air linked to 4 million deaths per year;
- if they know that the agreement includes a transfer $100 billion per year from the rich to the poor
- whether it is ethical to convert food to fuel to burn in their vehicles when it emits the same amount of greenhouse gases and drives up food prices disproportionately for the poor.

When they call you a conservative and a denier, correct them by explaining that you are a heretic, and that you are protecting the liberal values of the Age of Enlightenment. Inform them that extreme weather and wildfire events have not changed, so it's not you that is in denial. Ask them why the cod disappeared on the Grand Banks or the bison on the Great Plains. Was that global warming too, or are humans just too efficient at killing things? Why are the jungles being set on fire? Maybe ask them to stop scaring the children.

When they ask you what you are going to do to protect the next generation from climate change, tell them you plan on following the only policy that has worked for the entire co-existence of Homo sapiens and climate change: you will adapt.

And ask them to face the sun, close their eyes, and describe to you what they feel on their face. Then tell them we need that same sunlight to illuminate the climate debate. Dare them to know and to have the courage to use their own understanding.

Acknowledgements

Jane Everett: for planting the writing seed, nourishing it with thoughtful suggestions, and harvesting it with expert editing.

Josef and Patty Schachter, Dr. Carl Hodges: for honest and supportive critiques, and motivation to finish.

Lysle Barmby: for being my initial drafts reader and enabler, even though you had to.

CC: for being my illustration drafter, even though as a millennial global warmer, you didn't want to be associated with the book.

Terry Barmby: for your tireless and gifted copy editing and the deciphering of the language of engineering.

And to **Colonel L. Scott Loch, USMC:** for your kind comments that gave me the confidence to write this book.

List of Figures

Figure #1: <u>Comparison of a Greenhouse with the Greenhouse Gas Effect.</u> Original by Ron Barmby.

Figure #2: <u>Fish Weir Diagram.</u> Original by Ron Barmby.

Figure #3: <u>Logarithmic Effect of CO2 Sensitivity.</u> Original by Ron Barmby.

Figure #4: <u>Cross Section of the Sun.</u> Original by Ron Barmby.

Figure #5: <u>Sunspot Histogram Since 1700.</u> Originator: Royal Observatory of Belgium.

Figure #6: <u>Temperature Graph from Greenland Ice Cores.</u> Sourced from www.C3headlines.com and www.globalresearch.ca.

Figure #7: <u>The Second Law of Thermodynamics.</u> Original by Ron Barmby.

Figure #8: <u>Where is the Climate Emergency?</u> Original by Ron Barmby. Data from IPPC CO2 Scenario Processes for AR5 Recent Increases and Projected Changes in CO2 Concentrations.

Figure #9: <u>Is This the Climate Emergency?</u> Original by Ron Barmby. Data from www.metoffice.gov.uk/hadobs/hadcrut4.

Figure #10: <u>IPCC in 1990; Nothing to See Here, Folks.</u> Originator: IPCC FAR WG1 1990.

Figure #11: <u>If the Data Don't Fit, Hide it!</u> Original Ron Barmby. Data from IPCC TAR WG1. Many thanks to www.climatediscussionnexus.com video *Hide the Decline.*

Figure #12: <u>Frozen Greenhouse Gas (Methane) Bubbles in Abraham Lake.</u> Photo by Lysle Barmby.

Figure #13: <u>Walter Barmby and Family, circa 1934, Lang, Saskatchewan.</u> Photo source unknown.

Cover art: Jane Everett

Author photo: Lysle Barmby

Bibliography

Christian, David. *Origin Story: A Big History of Everything.* Little, Brown Spark; New York, 2018.

Hawking, Stephen. *A Brief History of Time: From the Big Bang to Black Holes.* Bantam Books; New York, 1988.

Intergovernmental Panel on Climate Change. *First Assessment Report, Climate Change: The IPCC Scientific Assessment.* 1990.

Intergovernmental Panel on Climate Change. *IPCC Second Assessment Report Climate Change.* 1995.

Intergovernmental Panel on Climate Change. *IPCC Third Assessment Report Climate Change.* 2001.

Intergovernmental Panel on Climate Change. *IPCC Fourth Assessment Report Climate Change.* 2007.

Intergovernmental Panel on Climate Change. *IPCC Fifth Assessment Report Climate Change.* 2014.

Intergovernmental Panel on Climate Change. *2019 Refinement to the 2006 IPCC Guidelines for National Greenhouse Gas Inventories.*

Issacson, Walter. *Einstein: His Life and Universe.* Simon & Shuster; New York, 2007.

Moran, Alan. *Climate Change: The Facts.* Stockade Books; Woodsville, New Hampshire, 2015.

Plimner Ian. *Heaven and Earth: Global Warming, the Missing Science.* Taylor Trade; Lanham Maryland, 2009.

Robson,   John.   *Climate   Discussion   Nexus.* climatediscussionnexus.com.

Wade, Nicholas. *Before the Dawn: Recovering the Lost History of Our Ancestors.* The Penguin Press; New York, 2006.

Walter, Chip. *Last Ape Standing: The Seven-Million-Year Story of How and Why We Survived.* Walker & Company; New York, 2013.

Watts, Anthony. *Watts Up with That.* wattsupwiththat.com.

About the Author

Ron Barmby is a Professional Engineer with both Bachelor's and Master's Degrees. His career has taken him to over 40 countries on five continents, studying local geosciences. His latest adventures include running the Boston marathon twice, climbing Kilimanjaro, and crossing Canada by motorcycle. Ron and his wife, Lysle, divide their time between Celista, British Columbia and Calgary, Alberta.